T0258541

Water Management: Insights an

# Agricultural Water Management: Insights and Challenges

Edited by **Keith Wheatley**

New York

Published by Callisto Reference,
106 Park Avenue, Suite 200,
New York, NY 10016, USA
www.callistoreference.com

**Agricultural Water Management: Insights and Challenges**
Edited by Keith Wheatley

International Standard Book Number: 978-1-63239-059-2 (Hardback)

Printed in the United States of America.

# Contents

# Preface

Various perspectives and challenges in agricultural water management have been discussed in detail in this book. Food security came up as a problem in the first decade of the 21st century, questioning the sustainability of humankind, which is certainly associated directly to the agricultural water management that has varied dimensions and needs integrative expertise in order to be dealt with. The aim of this book is to integrate the subject matter that deals with the modeling, auditing and assessment technique; and profitability, equity and irrigation water pricing in a single text. The book serves as a descriptive reference for professionals, students and researchers working on distinct aspects of agricultural water management. It is a comprehensive compilation of information regarding content revealing situations from distinct continents like USA, Europe, Africa, Australia and Asia. Several case studies have been elucidated in this book to provide the readers with a general scenario of the problems, challenges and perspectives on irrigation water usage.

The researches compiled throughout the book are authentic and of high quality, combining several disciplines and from very diverse regions from around the world. Drawing on the contributions of many researchers from diverse countries, the book's objective is to provide the readers with the latest achievements in the area of research. This book will surely be a source of knowledge to all interested and researching the field.

In the end, I would like to express my deep sense of gratitude to all the authors for meeting the set deadlines in completing and submitting their research chapters. I would also like to thank the publisher for the support offered to us throughout the course of the book. Finally, I extend my sincere thanks to my family for being a constant source of inspiration and encouragement.

**Editor**

# Part 1

# Equity, Profitability and Irrigation Water Pricing

# Irrigation Water: Alternative Pricing Schemes Under Uncertain Climatic Conditions

Gabriele Dono and Luca Giraldo
*University of Tuscia, Viterbo*
*Italy*

## 1. Introduction

The European Water Framework Directive (European Union, 2000; herein, WFD) aims to protect the environmental quality of water and encourage its efficient use. The EU member states are required to implement effective water-management systems and appropriate pricing methods that ensure the adequate recovery of water costs. These directive also relates to the pricing of water for agriculture. However, a general framework specific methodologies used by each country to establish water tariffs is not yet available.

Furthermore, it appears that numerous exceptional rules of contexts prevent the adoption of uniform pricing guidelines even within individual countries (OECD, 2010).

In the past decade, various studies have focussed on the pricing of irrigation water. Albiac and Dinar (Albiac & Dinar, 2009) published an up-to-date review of approaches to the regulation of non-point-source pollution and irrigation technology as a means of achieving water conservation, and Molle and Berkoff (2007) performed a thorough analysis of pricing policies worldwide, touching on multiple aspects related to water policy reform, primarily in developing countries. Tsur and others (Tsur et al., 2004) presented a similarly wide-ranging analysis. Most of these studies based their conclusions on the results of numerical modelling and generally did not consider the uncertainties that farmers face in making decisions (Bazzani et al., 2005; Riesgo & Gómez-Limón, 2006; Bartolini et al., 2007; Berbel et al., 2007; Semaan et al., 2007; Dono, et al., 2010). However, uncertainty related to climate change is an important aspect of decision-making in the context of the management of agro-ecosystems and agricultural production. In this regard, process-based crop models, such as Environmental Policy Integrated Climate (EPIC) (Williams et al., 1989), have been widely used to simulate crop response to changing climate, addressing the problem of assessing the reliability of model-based estimates (Niu et al., 2009).

Climate change related to the atmospheric accumulation of greenhouse gases has the potential to affect regional water supplies (IPCC, 2007). In particular, the long-term scenarios calculated by most global and regional climate models depict a greater reduction in precipitation with decreasing latitude in the Mediterranean area (Meehl et al., 2007). This result is important because reduced water availability could result in heavily reduced net returns for farmers (Elbakidze, 2006).

There are various sources of uncertainty in climate change simulations (Raisanen, 2007), including those associated with the nature of the direct relationships between climate variability and water resources, given the strong influence on such relationships of land cover (Beguería et al., 2003; García-Ruiz et al., 2008) and water-management strategies (López-Moreno et al., 2007). The main problems for irrigation reservoirs are that they must be filled at the beginning of the irrigation season, whereas the filling season is characterized by a large uncertainty. Consequently, the management regimen of the reservoir, and even the pricing of its water resources, must be adjusted to the variable conditions of inflow.

## 2. Aim of the study

The present study assesses the economic effects and influence on water usage of two different methods for pricing irrigation water under conditions of uncertainty regarding the accumulation of water in a reservoir used for irrigation. For this purpose, several simulations are performed using a Discrete Stochastic Programming (DSP) model (Cocks, 1968; Rae, 1971a, 1971b; Apland et al., 1993; Calatrava et al., 2005; Iglesias et al., 2007). This type of model can be used to analyse some of the uncertainty aspects related to climate change (CC) because it describes the choices open to farmers during periods (stages) in which uncertainty regarding the state of nature influences their economic outcomes. The DSP model employed in this study represents a decision-making process based on two decisional stages and three states of nature, reflecting different levels of water accumulation in the reservoir (Jacquet et al., 1997; Hardaker et al., 2007; Dono & Mazzapicchio, 2010).

The model describes the irrigated agriculture of an area in North-Western Sardinia where water stored in a local reservoir is distributed to farmers by a water user association (WUA). Simulations are executed to evaluate the performances of the different water-pricing methods when the conditions of uncertainty regarding water accumulation in the dam are exacerbated by the effect of climate change on winter rainfall[1]. In fact, the model simulations are first executed in a present-day scenario that reproduces the conditions of rainfall and water accumulation in the dam during 2004[2]. The model is then run in a scenario of the near future, which is obtained by projecting to 2015 the rainfall trends of the last 40 years[3].

Among the various productive and economic impacts of the methods for water pricing that the WUA may apply, particular attention is paid to examining the changes in the extraction of groundwater from private wells in the various scenarios. This resource is used by farmers

---

[1] Rainfall is most abundant in winter, making this season the most important in determining the level of water accumulation in the dam.

[2] The present-day scenario focuses on 2004 because a detailed sequence of aerial photographs, showing land use in northwest Sardinia throughout the agricultural season, is available for this year, courtesy of the MONIDRI research project (Dono et al. 2008). These photographs enable us to evaluate the ability of the model to replicate the choices of farmers in terms of soil cultivation.

[3] We chose a near-term future scenario because the Italian agricultural policy barely extends beyond 2013, given the upcoming implementation of the Common Agricultural Policy. The climate scenario for this period will be crucial for farmers in terms of deciding to adhere to the RDP measures that support adaptation strategies to climate change. In addition, extrapolating trends to a longer-term climate results in greater uncertain regarding the quality of the climate scenario. Finally, a longer-term scenario would increase the likelihood that the farm typologies and production technologies considered in this study would have become completely obsolete.

to supplement dam water, and its over-extraction is a key issue of environmental protection in the Mediterranean context.

## 3. Background

### 3.1 Payment schemes

In Italy, irrigation water is distributed by local associations of farmers (WUA) that 'water storage and distribution facilities developed mainly using public funding. In line with the guidelines of the WFD, Italian WUAs charge the associated farmers for the operating costs of water distribution, the maintenance costs of water networks, and the fees paid to local authorities, representing the opportunity costs of water and the environmental costs of providing the water. This set of items is herein referred to as the cost of water distribution (WDC, Water Distribution Cost). In most cases, the water storage and distribution facilities were built with public money, meaning that their long-term costs (depreciation and interest) are not included in the budgets of the WUAs, which only manage the water distribution service. Consequently, these costs are not included in farmers' payments to the WUA for irrigation costs. Note that there is a recent trend for farmers to co-finance investments in irrigation infrastructure, in which case the farmers also bear the long-term costs in proportion to their participation.

WUAs adopt various methods for charging WDC, with the most widely used being a fee that is paid per irrigated hectare. Some WUAs levy a two-stage fee (binomial system).

The per-hectare fee has traditionally been the most widely adopted method in Italy because it is the simplest to manage in terms of charging farmers. In fact, WUAs compute WDC at the end of the irrigation campaign and divide it by the amount of farmland that water was supplied to by the collective irrigation network, regardless of whether the land was irrigated; consequently, this approach bears no relation to the amount of water used by farmers.

The two-stage system comprises a *basic payment* and a *water payment*. The *water payment*, directly or indirectly linked to water use, is computed by multiplying the unit price of water by the amount of water used by farms. *Water payments* that are directly linked to water use are calculated based on readings from water meters installed at farm gates, while those that are indirectly linked to water use are calculated by estimating the water needs (per hectare) for each irrigated crop. The unit price of water is usually defined before the beginning of the irrigation season and is generally set below the expected average WDC. Farmers are then asked for a *basic payment* which covers general and maintenance costs and that is usually charged to individual farms according to the area of land equipped with the collective irrigation network. The *water payment* component of this two-stage system can be calculated using two different methods.

A VPM (Volumetric Payment Method) approach is used in the case that water meters are installed and functioning on every farm (as this enables water use to be monitored). This approach does not usually apply when water is delivered to the farm gate by gravity-fed canal networks. National and Regional Governments commonly provide financial support to encourage a switch from canal to pipeline systems and to install farm-gate meters as part of collective networks. This financial support aims at reducing water losses from the network and providing a better service to farmers, but also at metering water supplied to farms and encouraging the switch to VPMs.

Alternatively, *water payments* are calculated using an Area-Based Pricing Method (ABM), which estimates the unitary irrigation requirements for each irrigated crop (i.e., crop-based charges). Some WUAs calculate large, accurate sets of estimates that vary according to crop type, irrigation technology, soil characteristics and climatic conditions. In contrast, other WUAs refer to broad groups of crops with different unitary irrigation requirements, although this approach yields only a rough estimate of farm water use. In the case that an ABM is employed, farmers must apply to the WUA for water by reporting their irrigation plan at the beginning of the season. The WUA then checks if the actual extent of irrigated crops is consistent with the irrigation plan (to prevent the avoidance of payments in the case that the plans show fewer crops than actually cultivated). In the event of severe drought, during which time farmers are forced to leave fields fallow, payments are calculated based on the actual extent of irrigated crops, not solely on the cultivation plan presented beforehand.

ABMs are based on irrigated acreage and the water needs of crops, irrespective of whether the water comes from a WUA network or from farm wells, thereby generating an indirect charging effect on groundwater. VPM is widely supported in technical and political debates because it directly links water payments to the amount of water delivered to farmers. However, for both pricing models, water charges are set by WUAs in order to recover the WDC. The use of the average cost in these calculations deviates from the prescription that a fully *efficient* allocation scheme for a scarce resource such as water should be based on balancing the marginal net benefits of its uses (Perman *et al.*, 2003). However, these methods of charging farmers, even if economically imperfect, are easily manageable by WUAs.

## 3.2 Study area

The study area covers the Cuga River basin in the Sassari district, northwest Sardinia (Italy), comprising 34,492 ha of farmland (Figure 1). On 21,043 ha of this area, around 2,900 farms receive water from the Nurra WUA, distributing the surface water stored in two man-made lakes, Cuga (30 million $m^3$) and Temo (54 million $m^3$).

The WUA distributes only surface water: groundwater is managed by farmers as a private asset. In this system, the water stored in the two lakes is shared between urban and farm uses. In the case of a water shortage, urban uses are given priority and farmers respond by using water from private wells, if available.

Surface water is distributed via two interconnected network systems that differ in altitude (i.e., for low and high land). For lowland areas, water from the two lakes is directly introduced into pipelines and distributed by gravity. For highland areas, water is first pumped into gathering basins located at a relatively high altitude, from where it flows downward under gravity through a network of pipelines. In 2004, the two systems carried similar volumes of water. The water fees paid by farmers are aimed at recovering the WDC incurred by the WUA. Since 2001, the pricing method has been VPM, whereby farmers pay 0.0301 €/$m^3$ (in 2004) as a *water payment* for the water they use, measured via farm-gate meters installed at each farm of the WUA. Before 2001, the Nurra WUA adopted an ABM based on per-hectare estimated water use for three different groups of crops (Table 1)[4].

---

[4] In the study area, water meters were installed at farm gates with a financial contribution from National and Regional Governments.

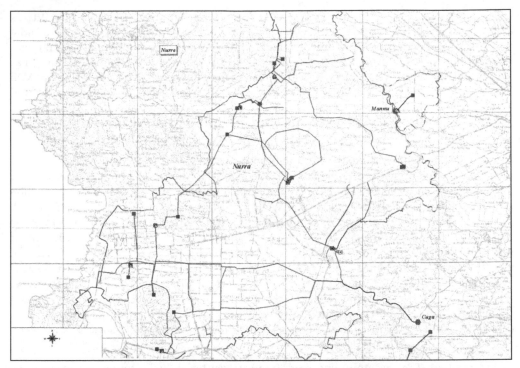

Fig. 1. WUA area of Nurra, North-Western Sardinia (Italy). Blue lines are the WUA boundaries; black line is the main channel from the reservoir to the pipeline network; redis the pipeline network.

| | Areas served by the WUA | | | | | Areas *not* served by the WUA | | | |
|---|---|---|---|---|---|---|---|---|---|
| | Well | No. of farms | Ha per farm | Cattle (heads) | Sheep (heads) | Well | No. of farms | Ha per farm | Cattle (heads) | Sheep (heads) |
| Cattle, L** | | 2 | 532.4 | 1,558 | - | | - | - | - | - |
| Cattle, S | | 5 | 37.6 | 280 | - | w | 27 | 12.6 | 1,026 | - |
| Crops, L | w* | 139 | 66.1 | - | - | w | 52 | 55.8 | - | - |
| Crops, M | w | 28 | 10.2 | - | - | w | 148 | 7.7 | - | - |
| Crops, S | | 1,509 | 0.7 | - | - | w | 540 | 0.9 | - | - |
| Olive, M | w | 33 | 12.3 | - | - | w | 25 | 13.4 | - | - |
| Olive, S | | 543 | 0.8 | - | - | w | 542 | 1.0 | - | - |
| Horticultural, M | w | 41 | 14.8 | - | - | w | 8 | 8.7 | - | - |
| Horticultural, S | w | 49 | 2.8 | - | - | w | 10 | 2.1 | - | - |
| Sheep, M | w | 34 | 64.1 | - | 57,578 | w | 33 | 76.3 | - | 14,398 |
| Sheep, S | w | 94 | 26.1 | - | 40,353 | w | 45 | 34.4 | - | 21,273 |
| Vineyards, L | w | 1 | 693.0 | - | - | | - | - | - | - |
| Vineyards, M | w | 136 | 2.7 | - | - | w | 44 | 2.0 | - | - |

* Farm possesses a private well.
** L, Large; M, Medium; S, Small.

Table 1. Farm typologies in the areas served and not served by the Nurra WUA.

There are no official data on the extraction of water from wells; however, the WUA's engineers have estimated that the annual withdrawal of groundwater is between 2.5 and 4 million m3, depending on how much water from the dams is provided for agricultural use. The number of wells owned by farms, as well as their location and technical features, has been identified from Agricultural Census data and from data compiled as part of the RIADE Research Project, jointly run by ENEA (National Agency for New Technology, Energy and Sustainable Economic Development, Italy) and the University of Sassari (Italy) (Dono et al., 2008). These data reveal that farms use approximately 107 wells in the area.

The agricultural sector of this territory is represented by a regional DSP model consisting of 24 blocks describing the most relevant farming systems. Each farming system, called a macro-farm (with reference to the block in the model), represents a group of farms that are homogeneous in terms of size (cultivated land and number of livestock head), production patterns, labour availability, presence of wells and location within the study area (Table 1). These macro-farms are defined using data from field surveys, the 2001 Agricultural Census and records of the European FADN (Farm Accountancy Data Network). The availability of multiple sources of farm data enabled us to consider economic characteristics (e.g., budget, net profit and performance indexes) in defining macro-farms. Thirteen of the macro-farms are located in the zone to which the WUA delivers water; 11 are located outside of this zone, where farms rely solely on water from privately owned wells or practice rain-fed agriculture. Note that the production of some of these typologies is not considered as typically Mediterranean, such as intensive dairy production and the associated cultivation of irrigated crops as forage.

In the mathematical programming model, production technologies for crops and livestock breeding are accurately defined based on the main activities observed in the study area. In particular, the use of water by crops is defined according to the employed irrigation techniques. Drip irrigation techniques, used for horticultural and tree crops, are represented in the model, whereas flood irrigation is not because this technology is not employed in the area. Farm typologies and production technologies that characterize the agricultural sector of the area were reconstructed as part of the MONIDRI Research Project, run by INEA (National Institute for Agricultural Economics, Italy) (Dono et al., 2008).

## 4. Methods

### 4.1 DSP models (general characteristics)

Discrete Stochastic Programming models (Cocks, 1968; Rae, 1971a, 1971b; Apland et al., 1993; Calatrava et al., 2005; Iglesias et al., 2007) can be used to analyse some of the uncertainty aspects related to CC. DSP models describe choices made by farmers during periods (stages) of uncertainty regarding conditions. Therefore, such models represent the decision process that prevails under typical agricultural conditions, where farmers are uncertain regarding which state of nature will prevail in the cropping season that is being planned, and it is only possible to estimate the probability distributions of the various states of nature. In this study, the DSP model represents a decision-making process based on two decisional stages and three states of nature (Jacquet et al., 1997; Hardaker et al., 2007), where farmers face uncertainty regarding the wintertime accumulation of water in a dam. In the literature, two-stage DSP models have considered various states of nature in the second stage. Jacquet and others (Jacquet et al., 1997) used four states of nature associated with annual rainfall, and Hardaker and others (Hardaker et al., 2004) represented the planning problems in dairy farming by referring to three levels of milk production. However, these authors did not justify the number of stages or the number of states of nature employed in the analyses, except for the need to simplify the problem as much as possible.

The first of the two stages of the DSP model proposed in this paper represents an autumnal period of choice, when farmers establish fields for winter crops. The limited irrigation needs of these cultivations can be satisfied by extraction from farm wells, and hence they are not directly influenced by uncertainty about water availability from the dam. However, when defining the area for winter crops, farmers also establish the surface to be left for spring crops. In contrast to winter crops, the irrigation needs of spring crops are substantial and can only be met by using water accumulated in the dam, whose availability is uncertain. In this way, uncertainty about water availability during the spring period influences the farmers' choices in the autumn period.

The second stage of the DSP model concerns the spring–summer period of choice. At that time, winter accumulation of water in the dam has already occurred, and farms can choose the area to be allocated to each spring crop with certainty. However, during this period farmers can only cultivate the area left unused from the first stage, when uncertainty about water levels in the dam might have produced choices that, in spring, turn out to be sub-optimal. This uncertainty is expressed by a probability distribution function of the level of water accumulation in the dam. The distribution is then discretized to yield three states of water accumulation (high, medium and low) along with their associated probability of realization.

The DSP model represents the influence of this uncertainty on the decision-making processes of farmers. According to this model, the farmer knows that different results may arise in planning the use of resources based on a certain state of nature. In particular, with three states of nature, three different results may occur. One is optimal, when the state of nature assumed by the farmer occurs as expected. The other two results are sub-optimal, where the farmer plans resource allocation based on a certain state of nature, but one of the other two states occurs, resulting in reduced income compared with the optimal outcome. The probabilities of these three results are the probabilities of the respective states of nature. The DSP represents the decision-making processes of the farmer who, based on these data, calculates the expected income of all the various outcomes (obtained by weighting the incomes from the three results with the probabilities of the respective states of nature) and adopts the solution that yields the higher expected income. Accordingly, the farmer adopts the use of resources generated by a weighted average of the three solutions.

Note that a solution that also weights the sub-optimal results may represent the outcome of precautionary behaviour of farmers who try to counter programming errors generated by relying on a given state of nature that ultimately does not occur. Also note that this average of DSP outcomes is different from the average of LP (Linear Programming) model outcomes under low, medium and high water-availability scenarios. Indeed, LP results are optimal to the relative water-availability state, considered in the LP model to be known with certainty. In contrast, DSP outcomes are sub-optimal when a state is planned but does not eventuate, meaning that average income levels are smaller than the analogous income levels in the LP model. This difference can be considered as the cost of uncertainty.

A major limitation of this approach may be that the farmer represented by the DSP model is risk-neutral; thus, the lower resulting income represents the cost of making optimal choices under conditions of uncertainty, but does not consider the cost of the farmer's attitude towards risk (risk aversion). Another limitation may be that we considered only one factor of uncertainty, whereas the farmer's decision-making process is affected by multiple uncertain factors that overlap. The future development of this analysis would be as a multi-stage DSP model with a larger number of uncertainty factors. However, with increasing number of stages and factors, the model becomes difficult to handle; consequently, it is crucial to identify the most relevant elements.

## 4.2 DSP model (technical characteristics)

As mentioned above, the DSP model used in this analysis is articulated in blocks of farm typologies. Each block refers to a macro-farm that represents a group of farms in the study area. The macro-farms differ in terms of structural characteristics (quality and availability of fixed resources in the short term), farming system and location. The optimisation problem involves maximising the sum of the stochastic objective functions of single macro-farms (expected gross margins), subject to all of the farming restraints (specific as well as territorial). Expected gross margins for each state of nature are given by the sum of two elements: one obtained from activities started in the first stage, and the other obtained from activities of the second stage. This DSP model can be mathematically formalised as follows:

Objective function:

$$\underset{X}{Max} \quad Z = GI_1 * X_1 + P_K * GI_2 * X_{2,K} \tag{1}$$

Subject to:

$$A_1 * X_1 + A_2 * X_{2,K} \leq b_K \quad \forall K \tag{2}$$

$$X_1, ..., X_{2,K} \geq 0 \quad \forall K \tag{3}$$

where $Z$ is the total gross margin, $X_1$ is the vector of first-stage activities, $X_{2,K}$ is the matrix of second-stage activities for each state of nature occurring in the second stage, $P_K$ is the probability of occurrence of each state of nature, $GI_1$ and $GI_2$ are the vectors of unitary gross margins, $A_1$ and $A_2$ are the matrixes of technical coefficients, $b_K$ is the vector of resource availability for each state of nature (here, only the availability of water has a different value for each state of nature; other resources have the same value), K is the state, and 1 and 2 are the stages. The variables of the model can be divided into three groups: crop, breeding and animal feeding; acquisition of external work; and activity related to the water resource.

Several groups of model constraints are defined. The first group refers to the expected availability of labour, land and water. Labour constraints are specified with reference to family labour and hired labour, permanent or temporary. Water constraints apply to both the reservoir water supplied by the WUA, as well as the groundwater, which can only be utilised based on the presence and technical characteristics of wells on the farms. The constraints on the expected availabilities of labour, land and water are specified for each month. Another set of constraints is concerned with agronomical practices as commonly adopted in the area to avoid declines in crop yields. Other constraints refer to Common Agricultural Policy systems to control production, such as production quotas and set-asides. Moreover, livestock breeding requires a balance between animal feeding needs and feed from crops or purchased on the market. In addition, constraints are imposed for specific farm typologies on the number of hectares of various trees growing and on the number of raised cattle or heads of sheep. These constraints are applied at different levels: some are specified at the farm level, such as the constraints on land use and on family and permanent labour, which cannot exceed the farm availability of these resources; others act at the area level, such as the constraint on the total irrigation water provided to farms, which cannot exceed the total water resources available to the WUA. Similarly, a constraint on temporary hired labour is specified at the area level. Finally, constraints on water availability are specified for each state of nature, for each of three scenarios regarding the distribution of water accumulation. Input and Produce Prices are defined as values that could be expected in 2004, based on the average of actualised values in the 3 preceding years. Similarly, agricultural policy conditions in 2004 are applied (Dono et al., 2008).

In essence, the basic approach of this study is to use a regional DSP model to estimate the impact of CC on production activity and income of farms in the area and to assess the performances of the various water-pricing methods under different climatic conditions. The stochastic expectations of water accumulation in the dam, which are included in the DSP model, are considered to be altered by CC that modifies the rainfall regime. The present-day (2004) probability distribution of water accumulation in the reservoir is estimated and used as a proxy for the stochastic expectation in the DSP model that reproduces the present conditions. This distribution is replaced with a future scenario probability function for rainfall and, hence, for the level of water accumulation in the dam. This future scenario is obtained by projecting historical rainfall data.

The next section describes the criteria used to reconstruct climate scenarios of winter precipitation and the resulting probability distributions for the accumulation of water in the dam, in the present and future.

## 4.3 Climatic scenarios

The present and future scenarios for water accumulation in the reservoir were reconstructed using the statistical correlation between rainfall amount and water storage in the dam, and by extending to the 2015 year the estimated trend of a 40-year rainfall series.

Estimation of the probability distribution for water accumulation in the reservoir was complicated by the fact that the Nurra WUA was only able to provide accurate monthly data for short periods in recent years. At the time of the MONIDRI research project, accurate records were only available for the years 1992–2003. Table 2 lists the annual values of water allocation obtained from these monthly data, showing that on average, potable use accounted for 40% of the available resource. In the years 1995, 2000 and 2002, the total amount of water available was insufficient to meet all the needs, and the Commissioner for Water Emergency limited the amount withdrawn for irrigation in favour of domestic usage, which had a major impact on farm incomes. During these years, the withdrawal for domestic use exceeded that for irrigation.

| Water uses | 1992 | 1993 | 1994 | 1995 | 1996 | 1997 | 1998 | 1999 | 2000 | 2001 | 2002 | 2003 |
|------------|------|------|------|------|------|------|------|------|------|------|------|------|
| *Irrigation* | *43.5* | *33.4* | *31.6* | *2.3* | *23.2* | *39.1* | *10.4* | *17.6* | *2.4* | *27.8* | *12.4* | *26.6* |
| *Potable* | *10.3* | *11.2* | *11.3* | *10.9* | *12.8* | *12.7* | *13.2* | *14* | *14.3* | *20.4* | *19.7* | *8.1* |

Table 2. Amount of water (million m³) from Cuga Dam allocated to different uses in the period 1992-2003 (source: Nurra WUA).

The limited temporal coverage of this record of water use makes it statistically insufficient for estimating the probability distribution of states of water storage for the present scenario, and even more so for the future. In addition, hydrological models had not been developed for the study area for the appropriate transformation of long-period rainfall data in terms of water accumulation in the Cuga Dam. To overcome these limitations, a statistical relationship was estimated between rainfall amount and water accumulation level, and the parameters of this relationship were used to generate the probability distributions of water collection states. The following section describes the procedure for estimating the statistical relationship between rainfall regime and level of water accumulation in the reservoir. These estimated values are used to obtain the probability distribution of water level in the reservoir, for which low, medium and high states of accumulation were defined.

### 4.4 Assessment of climate change

The first step in the analysis was to examine the long-term trends in the rainfall regime that are believed to have influenced the accumulation of water in the Cuga Dam. Rainfall in the area was analysed using a 43-year series of monthly data (1961–2003) comprising a total of 516 observations. This analysis assumed an additive or multiplicative relationship between the components. The choice between additive or multiplicative decomposition methods was based on the degree of success achieved by their application (Spiegel, 1973). In this study, the multiplicative method yielded slightly better results than the additive decomposition. The analysis was therefore based on the assumption that the following multiplicative link exists among components:

$$X = T * S * C * \varepsilon \tag{4}$$

where X is the observed rainfall data as generated by trend T, seasonality S, cycle C and residual elements ε. The influence of these elements was decomposed. To estimate the trend, a linear function was used as follows:

$$\text{Rain} = \delta_0 + \delta_1 T + \varepsilon \tag{5}$$

where Rain is rainfall, T is time and $\delta_0$ and $\delta_1$ are the parameters of the function. Quadratic or exponential functions can also be used for estimating trends; the choice among the different structures is generally based on their statistical adaptation to the analysed series (Levine et al., 2000).

Seasonality (S), as a specific characteristic of each individual month, was obtained by first normalising the monthly data to the average for that year, and then computing from these values the median for each month in the observed range. We assumed the absence of a cycle (C) in climatic events of the study area, given the lack of clear physical phenomenon (e.g., a dominant atmospheric circulation pattern) linked to cyclic behaviour in the study area.

Finally, residuals were calculated by isolating the observed data from the climatic components of trend and seasonality, given that any cycle is assumed to be absent. Residuals usually depend solely on random and uncertain factors; i.e., they are stochastic elements that represent the variability of climate phenomena. Analysing the standard deviation of residuals can highlight the existence of temporal changes in the variability of climate phenomena, which is an important part of CC. This analysis can be developed by estimating a linear trend of the standard deviation of residuals, as follows:

$$SDR = \gamma_0 + \gamma_1 T + \varepsilon \qquad (6)$$

where SDR is the standard deviation of residuals and $\gamma_0$ and $\gamma_1$ are the parameters of the function.

## 4.5 Statistical relation between rainfall regime and water accumulation in the Cuga Dam

Once the rainfall data had been examined and the presence of relevant trends highlighted, a statistical relationship was estimated to define the on water accumulation in the Cuga Dam. A linear regression model between rainfall (Rain) and water amount in the dam (Wa) was constructed based on the 144 monthly observations for the period 1992–2003. The estimated coefficients of this regression and the observed rainfall data were used to reproduce the entire series of data on water amount in the dam for the period 1961–2003. This procedure generated a sufficiently long series of data on water accumulation in the dam to be used when estimating the probability distribution of this variable.

In more detail, when estimating the statistical relationship between rainfall data and water accumulation level, a preliminary analysis of the data was performed to reveal (and eventually correct) the possible non-normality and non-stationarity characteristics of the series. A series is considered normal when the characteristics of symmetry and unimodality make it similar to the realisations of a normal random variable. A stationary process is a stochastic process whose joint density distribution does not change when shifted in time or space; as a result, parameters such as mean and variance (if they exist) also do not change over time or space. To satisfy the regression model hypothesis, data that do not show normality or stationarity characteristics must be standardised to obtain a stationary series, as follows:

standardisation:

$$X_{SAV_{i,j}} = \frac{X_{OBS_{i,j}} - \mu_i}{\sigma_i} \qquad (7)$$

where $X_{SAVi,j}$ is the series of seasonally adjusted values, $X_{OBSi,j}$ is the series of observed values, $\mu_i$ is the monthly average of the values of the observed series, $\sigma_i$ is the monthly standard deviation of the observed series, and i and j indicate the month and year, respectively.

Therefore, the standardised data were normalised by using a Box–Cox transformation:

normalisation:

$$X_{tras_{i,j}} = \begin{cases} \dfrac{(X_{SAV_{i,j}})^{\lambda} - 1}{\lambda} & \text{for } \lambda \neq 0 \\[2mm] \log(X_{SAV_{i,j}}) & \text{for } \lambda = 0 \end{cases} \tag{8}$$

where $\lambda$ is determined by maximising the following log-likelihood function:

$$\ln L(\lambda, \overline{x}_{SAV}) = -\frac{n \cdot m}{2} \ln \left[ \sum_{i=1}^{n} \sum_{j=1}^{m} \frac{(x_{SAVi,j}(\lambda) - \overline{x}_{SAV}(\lambda))^2}{n \cdot m} \right] + (\lambda - 1) \sum_{n}^{i=1} \sum_{j=1}^{m} \ln(x_{SAVi,j}) \tag{9}$$

and

$$\overline{x}_{SAV}(\lambda) = \begin{cases} \dfrac{1}{n \cdot m} \sum_{i=1}^{n} \sum_{j=1}^{m} (x_{SAVi,j})^{\lambda} & \text{for } \lambda \neq 0 \\[3mm] \dfrac{1}{n \cdot m} \sum_{i=1}^{n} \sum_{j=1}^{m} \ln(x_{SAVi,j}) & \text{for } \lambda = 0 \end{cases} \tag{10}$$

where n and m are respectively the number of months and years.

Hence, making use of standardised and normalised data on water amount in the dam (Wa$_{tras}$) and on rainfall (Rain$_{tras}$), we constructed the following model:

$$Wa_{tras} = \beta_0 + \beta_1 Rain_{tras\,(-1)} + \beta_2 Wa_{tras\,(-1)} + \varepsilon \tag{11}$$

Based on the coefficients of this model, Wa$_{tras}$ values were first calculated for the period 1961–1991 and then transformed into water availability data (Wa) by applying the inverse Box–Cox transformation (inverse of normalisation) and the inverse standardisation adjustment.

## 4.6 Probability distribution of water accumulation in the dam

The inferred data on water accumulation levels for individual months during the period 1961–2003 were used to estimate density functions related to sub-periods within this interval. These functions were estimated on the basis of a dataset restricted to March values, because this month is the last before the start of the irrigation campaign. Consequently, the level of water accumulation in March is crucial for decisions made by farmers regarding the cultivation of irrigated and non-irrigated crops, and for decisions made by WUAs regarding water allocation to farms. The parameters of these density functions are estimated using the software @Risk, which uses a chi-square value (Goodness of Fit index) to select the function that best approximates the dataset. These parameters are the basic input for generating the stochastic expectations of farmers in the DSP model.

Probability values can be computed and incorporated into DSP models only for states of nature expressed as intervals and not as single values (Piccolo, 2000). To this end, based on water management in the area of interest, three accumulation states are considered as relevant to the use of reservoir water in the farm sector: high, medium and low. The first state is recognised as occurring in years when the dam contains abundant water, when no limits are imposed on water use for irrigation or other purposes. A state of medium accumulation is identified for years when the amount of collected water necessitates careful use, even if explicit measures of public rationing are not required. A state of low accumulation is recognised when major water emergencies occur and irrigation is limited by public authorities to ensure the availability of water for potable use. The boundaries between these states are defined based on their occurrence in 1992–2003. Specifically, the lower limit of the low accumulation state is taken as the minimum value of the series: 5.6 million cubic meters ($Mm^3$). The upper bound of this state is taken as the maximum value at which irrigation was publicly rationed to guarantee potable use (42.6 $Mm^3$). This latter value is also the lower bound of the medium state, whose upper limit (64.0 $Mm^3$) is defined based on symmetry about the average value of accumulation (53.3 $Mm^3$). The value of 64.0 $Mm^3$ is also the lower bound of the high accumulation state, whose upper limit is the maximum value of the series, 89.9 $Mm^3$. Different sub-periods during the interval 1961–2003 yield different distributions of water accumulation states, with different probabilities and average values for the three states. The parameters of these different functions can be used to generate stochastic expectations of water accumulation in the dam, as represented by the regional DSP model.

The dataset obtained using this adjustment to the regression results was used to estimate a first probability distribution function for the continuous, stochastic variable of water accumulation in the dam, based on data for the years 1984–2003; this represents the expectations of the present period. Similarly, 21 distributions were computed by progressively shifting the 20-year period forward, by 1 year at a time, from the period 1964–1983 to the period 1984–2003. The probability values for low, medium and high levels of water accumulation in the dam were computed for each of these distributions. Based on the result, linear trends of probability values for the three states of nature were estimated and projected to estimate data for the years 2004–2015 and to compute an analogous probability distribution for the 20-year period 1996-2015. The probability distributions obtained in this way for the years 1984–2003 and 1996-2015 were used to represent the stochastic expectations in the present and future, respectively.

## 4.7 Simulation scenarios

The baseline model of this study refers to the VPM, as applied by the Nurra WUA in 2004, when the *water payment* was set at 0.0301 €/m³ for water delivered by the WUA distribution network and the *complementary payment* was charged to fully cover the WDC. With this pricing method, only the *water payment* directly affects farmers' water use. No charge for the use of groundwater is applied by the WUA.

Two other scenarios refer to ABM. In these cases, the farm payment for water consists of two components: the *water payment* is charged according to *estimates* of the water requirements of crops, multiplied by the water unitary price (0.0301 €/m³), and the *complementary payment* is again charged to ensure that the WDC is fully recovered.

Two ABM scenarios are simulated (ABM-1 and ABM-2), referring to two different methods of estimating the water applied to each crop. In ABM-1, these estimates accurately reflect the irrigation requirements of every crop, whereas in ABM-2 the crops are clustered into classes that consider their *average* irrigation requirements[5]. This latter scenario considers the ABM practiced by the Nurra WUA until 2001, where the estimates of unitary irrigation requirements used in calculating the payments are not always consistent with the actual requirements of crops (Table 3)[6]. The main feature of ABMs is that the payment is charged regardless of the water source (i.e., surface water from the WUA or groundwater from private wells). Furthermore, all irrigated areas are supposedly charged by the WUA. Given that farmers pay according to the area under irrigation and that the price is set considering estimates of irrigation requirements, irrigation payments are affected by cropping patterns but not by the source of water.

| I – (104.30€/ha) | II – (143.84 €/ha) | III – (179.77 €/ha) |
| --- | --- | --- |
| Tomatoes in glasshouses | Ryegrass | Artichoke |
| Watermelon | Alfalfa | |
| Melon | Clover | |
| Olive trees | Corn | |
| Vineyard | Open field Tomatoes | |
| Peach trees | | |

Table 3. Payment classes in the Nurra WUA (based on the parameters applied in 2001)

## 5. Results

First, we present the temporal changes in rainfall patterns over the past 40 years, and then describe the outcomes obtained by estimating a statistical relationship between the rainfall regime and level of water accumulation in the dam. Subsequently, the levels and respective probability values for states of water accumulation in the dam are reported for each of the three scenarios. Finally, we present the economic and productive outcomes of the DSP models.

### 5.1 Precipitation time series

The linear trend of monthly rainfall reveals a decrease in the area, indicated by the value of the regression coefficient in relation to time ($\delta_1$) (Table 4). In addition, the linear trend in the monthly variability of the standard deviations of residuals reveals an increase residuals, as indicated by the regression coefficient of the same standard deviations of residuals in relation to time ($\gamma_1$).

### 5.2 The regression model

A preliminary analysis of the data reveals a statistically significant autocorrelation in the Wa series and cross-correlation between the Wa and Rain series. The data were then

---

[5] Some WUAs chose this last option to reduce their administrative burden or because of political reasons, such as considering the relative contributions of certain crops in terms of farm employment or income.

[6] For instance, clover and alfalfa in the second class (Table 3) pay less than artichoke in the third class, yet the water needs of the former are approximately twice those of the latter. This favourable treatment is justified by WUAs because alfalfa and clover crops are relevant to the cow- and sheep-milk sectors, which are considered to be important for the economy of the entire area.

standardised and normalised, and new time series of the amount of water in the dam ($Wa_{tras}$) and rainfall ($Rain_{tras}$) were used to estimate the model (11).

|  | Coefficient | Estimate | T-stat | P value |
|---|---|---|---|---|
| Rainfall | $\delta_0$ | 55.645 | 13.6477 | 0.000 |
|  | $\delta_1$ | -0.0381 | -2.7904 | 0.000 |
|  |  |  |  |  |
| Standard deviation | $\gamma_0$ | 131.07 | 3.2885 | 0.000 |
| of residuals | $\gamma_1$ | 2.74 | 1.6568 | 0.100 |

Table 4. General trends of rainfall and of the standard deviation of residuals.

Good statistical results were obtained from the regression: the coefficients have the expected signs and are statistically significant, and high $R^2$ values indicate that more than 95% of the variability in $Wa_{tras}$ is explained by the model (see Table 5).

| Coefficient | Estimate | T-stat | P-value |
|---|---|---|---|
| $\beta 0$ | -0.3149 | 4.5442 | 0.000 |
| $\beta 1$ | 0.5828 | 5.3247 | 0.000 |
| $\beta 2$ | 0.9694 | 54.6706 | 0.000 |
|  |  |  |  |
| $R^2$ | 0.9555 | F | 1501.44 |
| Adj-$R^2$ | 0.9548 | $p \, (F \leq f)$ | 0.000 |

Table 5. Results of the regression model for the dependent variable $Wa_{tras}$.

Based on these coefficients, $Wa_{tras}$ values were calculated for the period 1961–1991 and were then transformed to water availability data (Wa) by applying the inverse Box–Cox transformation (inverse of normalisation) and the inverse standardisation adjustment.

## 5.3 The density distribution of water accumulation

The dataset obtained from this adjustment of the regression results was used to estimate the density functions of water accumulation in the dam for the period 1984-2003. The best estimate was a triangular function with a chi-square value of 0.400 and a p-value of 0.9402. With k (number of bins) = 4 (3 degrees of freedom), the null hypothesis is accepted (i.e., that this is the best possible function for representing the data). Once the boundaries of the low, medium and high accumulation states were defined based on data for the period 1992–2003, the respective probabilities were computed, yielding values of 27.3%, 40.7% and 32.0%, respectively.

Next, the density distribution values for the future scenario were determined. To this end, 21 distributions were computed by progressively shifting the 20-year window, 1 year at a time, from 1964–1983 to 1984–2003. The probability values for low, medium and high levels of water accumulation in the dam were computed, yielding a progressive increase in the degree of variability, especially for periods more recent than 1976–1995. Based on these data, linear trends of probability values for the three states of nature were estimated and projected to obtain an analogous distribution for 1996–2015, yielding values of 38.8%, 13.7% and 47.5% for the states of low, medium and high water accumulation in the dam, respectively.

Once the total accumulation levels and respective probabilities had been defined for the two scenarios, the respective availabilities of water for irrigation were also defined. To this end, the data supplied by the WUA for the years 1992–2003 were used to infer the level of water accumulation in the dam and the percentage of water allocated to irrigation for each of the three states of nature. These percentages were used to define the amount of water accumulated in the dam that farmers could expect to be allocated to agriculture in each state of nature.

As a side analysis, an analogous distribution based on data for the period 1964–1983 was computed with the same boundaries for water collection states, yielding probability values of 0.0%, 99.7% and 0.3% for the low, medium and high levels, respectively. Compared with the previous scenario, these outcomes reveal that water accumulation in the dam during the 1960s and 1970s was characterised by a smaller variability than in present scenario. This result is consistent with the finding of a temporal increase in rainfall variability, as obtained by analysing the standard deviation of residuals (see Table 4).

## 5.4 The DSP models

The DSP simulation models employed in this study were solved using the program GAMS (General Algebraic Modelling System; Brooke et al., 1996). The baseline of this study is the average of the outcomes related to the three states of nature in the present scenario, weighted by the respective probabilities. The baseline is evaluated by comparing its average weighted outcome regarding land use to the actual pattern determined from remote sensing data and field data approved by the WUA (Dono et al., 2008). The similarity between the patterns is assessed by the Finger–Kreinin index, which compares the respective percentages of total land occupied by each group of crops, selects the lower value among them and sums the figures (Finger and Kreinin, 1979). The more similar that two series are, the higher the sum of the lower values, which yields a value of 100% for identical series. A high degree of similarity is obtained in this study between DSP outcome and the actual land use, with the value of the similarity index being 91.9%. The baseline model is therefore considered to adequately reproduce the observed choices of farmers and is therefore useful for providing insights into farmers' possible adjustments in the case of changing economic or climatic conditions[7].

At this point, we can discuss the results obtained with different pricing methods for the water distributed by the WUA, in the context of the present and future climate. Table 6 lists

---

[7] An analogous linear programming model (LP) was constructed, differing only in the condition of irrigation water availability, which was defined as the average water level in the dam over the previous 5 years. By considering uncertainty, the DSP model yields better results in reproducing agricultural activities; in fact, the Finger–Kreinin similarity index has a lower value (90.2) in the LP model.

the key financial results for the entire area in which the WUA distributes water from the reservoir.

Outside of this area, agriculture is not irrigated and is not affected by the water pricing system of the dam or changes in the volume of water in the reservoir. Table 6 lists the total revenues, indicating the portion of product sales, the main items of variable costs and gross margin. Fixed costs have been estimated for the various farm typologies, enabling the calculation of their net incomes; these are aggregated to yield the entire area value. Table 6 lists all of these values, expressed in thousands of euros and in percentage change from the baseline (in this VPM).

| | Absolute value ('000 €) | | | | | | % variations from the Baseline | | | | |
| --- | --- | --- | --- | --- | --- | --- | --- | --- | --- | --- | --- |
| | present | Present | | future | | | present | | future | | |
| | VPM | ABM | | VPM | ABM | | ABM | | VPM | ABM | |
| | Baseline | 1 | 2 | | 1 | 2 | 1 | 2 | | 1 | 2 |
| Revenue | 73,892 | 73,892 | 73,906 | 73,713 | 73,713 | 73,644 | 0.0 | 0.0 | -0.2 | -0.2 | -0.3 |
| Sales | 64,667 | 64,667 | 64,68 | 64,557 | 64,557 | 64,478 | 0.0 | 0.0 | -0.2 | -0.2 | -0.3 |
| Costs | 19,109 | 19,151 | 19,218 | 19,431 | 19,483 | 19,505 | 0.6 | 0.2 | 1.7 | 2.0 | 2.1 |
| Feeding cost | 430 | 430 | 430 | 723 | 723 | 723 | 0.0 | 0.0 | 68.2 | 68.2 | 68.2 |
| Labour cost | 2,133 | 2,133 | 2,134 | 2,34 | 2,34 | 2,321 | 0.1 | 0.0 | 9.7 | 9.7 | 8.8 |
| WUA cost | 341 | 407 | 461 | 289 | 362 | 450 | 19 | 35 | -15 | 6 | 32 |
| Drawing cost | 82 | 58 | 58 | 88 | 67 | 66 | -29.7 | -29.7 | 6.8 | -18.7 | -19.1 |
| Irrigation Equip. | 1,381 | 1,381 | 1,381 | 1,379 | 1,379 | 1,365 | 0.0 | 0.0 | -0.1 | -0.1 | -1.2 |
| Other costs | 14,06 | 13,928 | 13,831 | 14,034 | 13,888 | 13,681 | -0.9 | -1.6 | -0.2 | -1.2 | -2.7 |
| Gross Margin | 54,783 | 54,742 | 54,688 | 54,283 | 54,231 | 54,139 | -0.2 | -0.1 | -0.9 | -1.0 | -1.2 |
| Net Income[8] | 32,627 | 32,532 | 32,586 | 32,127 | 31,983 | 32,075 | -0.3 | -0.1 | -1.5 | -2.0 | -1.7 |

Table 6. Economic results for the entire area.

These data show that the transition from VPM to the ABMs generates a very small change in income in the present climate scenario. However, a significant change in cost structure emerges, with a strong reduction in expenses for the extraction of water from wells and an increase in the irrigation payments to the WUA. This change occurs because the two ABMs are based on irrigated acreage and the water needs of crops, irrespective of whether the water is derived from the WUA network or from farm wells, thereby generating an indirect pricing of groundwater. Consequently, the two ABMs encourage farmers to reduce the use of groundwater that is only applied in cases where irrigation is necessary but the WUA irrigation season has yet to open, or during the summer periods when the water resources of the WUA do not meet the general water demand of the area.

---

[8] Net income is obtained based on estimates of fixed costs coming from European FADN database.

The transition to the future climate scenario results in a more pronounced change in cost structure. The greater variability in water accumulation in the reservoir generates a greater reduction in total income, which also affects the system based on VPM. With this method of charging, there is an increase in the cost of drawing groundwater and a reduction in irrigation payments to the WUA. At the same time, there is an increase in the cost of purchasing feed and forage that can no longer be sufficiently produced locally under the new scenario of expectations regarding the availability of water in the reservoir. Of note, the use of ABMs yields the same increase in costs as when using VPM. However, these last two methods of water pricing are completely different from VPM in terms of the effect on other cost items. As in the present scenario, a reduction in expenses occurs with the extraction of water from wells, with a parallel increase in irrigation payments to the WUA. These variations are less pronounced than in the present scenario because the expectation of a greater variability in the future accumulation of water in the reservoir prevents a more significant reduction in the use of groundwater.

Table 8 lists the net incomes of the farm typologies in the study area for each scenario, grouped by product specialization. These data show that the farms involved mainly in the production of crops (cereals, oilseeds and protein crops, and also forage and pasture) make the greatest contribution (30%) to the agricultural income of the territory, followed by vineyards and to a much lesser degree the sheep farms and the other typologies.

In the case of ABM-2, the horticultural farms show a marked increase in income, which is not seen in the case of ABM-1. This result indicates that estimates of the water needs of the WUA in ABM-2 favour some vegetables grown by horticultural farms.

| Farm Typology | Absolute value ('000 €) | | | | | | % variations on the baseline | | | | |
| | present | Present | | future | | | present | | future | | |
| | VPM | ABM | | VPM | ABM | | ABM | | VPM | ABM | |
| | Baseline | 1 | 2 | | 1 | 2 | 1 | 2 | | 1 | 2 |
| Cattle | 1,363 | 1,364 | 1,365 | 1,207 | 1,208 | 1,206 | 0.1 | 0.1 | -11.5 | -11.4 | -11.5 |
| Arable | 13,738 | 13,691 | 13,634 | 13,381 | 13,339 | 13,322 | -0.3 | -0.8 | -2.6 | -2.9 | -3 |
| Olive | 3,525 | 3,54 | 3,545 | 3,511 | 3,511 | 3,529 | 0.4 | 0.6 | -0.4 | -0.4 | 0.1 |
| Vegetable | 982 | 982 | 1,048 | 1,050 | 1,050 | 1,029 | 0 | 6.7 | 6.9 | 6.9 | 4.7 |
| Sheep | 2,691 | 2,683 | 2,684 | 2,657 | 2,65 | 2,649 | -0.3 | -0.2 | -1.2 | -1.5 | -1.5 |
| Vineyard | 9,36 | 9,356 | 9,287 | 9,353 | 9,348 | 9,279 | 0 | -0.8 | -0.1 | -0.1 | -0.9 |
| Total | 31,658 | 31,617 | 31,563 | 31,158 | 31,106 | 31,014 | -0.1 | -0.3 | -1.6 | -1.7 | -2 |

Table 7. Net income of farming typologies.

However, the most interesting aspect of this table is the effect of the transition to the future scenario. The mean changes emerge as the result of very different changes among the typologies, with a collapse in the incomes of dairy farms and an appreciable increase in the income of vegetable growers. The choice of pricing system of irrigation water has little influence on the effect of increased variability in water accumulation in the dam.

The VPM scheme encourages farmers to meet their water requirements with minimum cost. Consequently, farmers with wells tend to draw water until it remains cheaper than the WUA *water payment*. Thus, the VPM scheme results in increased groundwater extraction. Under the ABM scenarios for the present, in contrast, farmers only draw groundwater from wells if the WUA is unable to supply water, either because of a demand pick or a request coming off the irrigation season. Otherwise, farmers are inclined to use the WUA water as much as possible, because such behaviour does not affect the *water payment*. When these pricing method are simulated in the future climate scenario, the amount of groundwater extraction is lower than that of the present in the case of the ABMs; moreover, the use of VPM results in increased extraction in the case of increasing uncertainty regarding the availability of WUA water.

| | Absolute value (ha) | | | | | | % variations on the baseline | | | | |
| | present | Present | | future | | | present | | future | | |
| Cultivation | VPM | ABM | | VPM | ABM | | ABM | | VPM | ABM | |
| | Baseline | 1 | 2 | | 1 | 2 | 1 | 2 | | 1 | 2 |
| Forage | 8,935 | 8,935 | 8,925 | 9,050 | 9,050 | 9,016 | 0 | -0.1 | 1.3 | 1.3 | 0.9 |
| Wheat | 3,679 | 3,679 | 3,679 | 3,771 | 3,771 | 3,771 | 0 | 0 | 2.5 | 2.5 | 2.5 |
| Barley/Oat | 850 | 850 | 850 | 869 | 869 | 888 | 0 | 0 | 2.2 | 2.2 | 4.5 |
| Pasture | 3,209 | 3,209 | 3,209 | 3,205 | 3,205 | 3,233 | 0 | 0 | -0.1 | -0.1 | 0.7 |
| Silage corn | 216 | 216 | 216 | 133 | 133 | 133 | 0 | 0 | -38.6 | -38.6 | -38.6 |
| Grain corn | 867 | 867 | 867 | 660 | 660 | 660 | 0 | 0 | -23.9 | -23.9 | -23.9 |
| Tomato | 21 | 21 | 21 | 21 | 21 | 21 | 0 | 0 | 0.0 | 0 | 0 |
| Artichoke | 243 | 243 | 253 | 83 | 83 | 85 | 0 | 4 | -65.9 | -65.9 | -65 |
| Melons | 1,061 | 1,061 | 1,061 | 1,148 | 1,148 | 1,133 | 0 | 0 | 8.2 | 8.2 | 6.8 |
| Olive | 754 | 754 | 754 | 754 | 754 | 754 | 0 | 0 | 0.0 | 0 | 0 |
| Wine | 1,336 | 1,336 | 1,336 | 1,336 | 1,336 | 1,336 | 0 | 0 | 0.0 | 0 | 0 |
| Peach | 587 | 587 | 587 | 587 | 587 | 587 | 0 | 0 | 0.0 | 0 | 0 |
| | 208 | 208 | 211 | 190 | 190 | 191 | 0 | 1.6 | -8.3 | -8.3 | -8 |
| Total | 21,966 | 21,966 | 21,970 | 21,807 | 21,808 | 21,807 | 0 | 0 | -0.7 | -0.7 | -0.7 |
| *Water sourcing* | | | | | | | | | | | |
| Total | 16,821 | 17,007 | 17,040 | 14,841 | 15,005 | 14,939 | 1.1 | 1.3 | -11.8 | -10.8 | -11.2 |
| WUA | 13,925 | 14,928 | 14,960 | 11,792 | 12,673 | 12,620 | 7.2 | 7.4 | -15.3 | -9 | -9.4 |
| Wells | 2,896 | 2,080 | 2,080 | 3,050 | 2,332 | 2,319 | -28.2 | -28.2 | 5.3 | -19.5 | -19.9 |

Table 8. Farming activities and water sourcing.

## 6. Discussion

The consequences of using different water-pricing systems for irrigation water were estimated by applying the systems under various climate scenarios (i.e., level of water accumulation in the reservoir). The first system is the one currently applied by the WUA, the Volumetric Pricing Method (VPM), based on the metered use of water by farms. The second system is an Area-Based Pricing Method (ABM), whereby fees are charged per hectare according to the estimated average water use for each crop. This system was applied in two versions: (1) employing water use coefficients that strictly reflect the actual irrigation requirements of the various crops in the area, and (2) employing the estimated average levels of water use prescribed by the WUA prior to 2001, when the switch was made to VPM. We used a DSP model to examine the application of these pricing methods in a future scenario in terms of their impacts on the use of agricultural land, on inputs (e.g., water and labour), and on the income of the agricultural sector, for the entire area and for representative farms.

The results of DSP modelling suggest that the farm sector overall is well placed to adapt to CC in the present and in the near future, particularly with respect to water accumulation in the dam. Indeed, the model predicts that increased variability in water accumulation in the reservoir would have a negligible effect on the economy of the entire agricultural sector in the study area. However, the economic impact of this increased variability shows marked differences among the farm typologies: some suffer marked reductions of income, particularly dairy farms that depend on the use of large volumes of water for the irrigation of corn.

Furthermore, the general adaptation path followed by the agricultural sector of the Nurra area is predicted to result in an increase in the environmental impact of agricultural activities, including the excessive extraction of groundwater. In this regard, the use of VPM poses problems when individual farmers have direct access to uncontrolled water sources such as groundwater, as is the case in the present study area and in many other Mediterranean areas. This problem arises because wells are generally a private asset of the farm and because there is a lack of information and legislation regarding this source of water, which would be required to control its level of exploitation. These results are consistent with the findings of other studies regarding the use of volumetric pricing (Cornish et al., 2004; Dinar et al., 1989). Furthermore, the application of ABM pricing, unlike VPM, is able to restrict the extraction of groundwater even in a scenario of increased uncertainty regarding water availability from collective, surface sources. This restriction arises because ABMs charge for the irrigation of crops regardless of the water source. Therefore, groundwater under VPM is a substitute for water distributed by the WUA, whereas under ABM it is complementary to the water distributed by the WUA, as its extraction generates extra pumping costs but does not save on other irrigation costs. This eliminates cost competition between the two water sources and results in a marked reduction in groundwater use.

These findings demonstrate that the introduction of VPMs is, in many regards, contradictory to the basic goal of environmental protection advanced by the WFD, since over-extraction could lead to increased salinization of groundwater. The pricing method can be considered a relevant strategy for adapting to the challenges of foreseen climate scenarios; hence, the adoption of a unique *a priori* strategy for water conservation may yield unsatisfactory results.

## 7. Conclusion

The contribution of this study is of interest primarily because it examines different pricing methods of irrigation water in a state of uncertainty, which is typical of the decision-making framework in the agricultural sector. Moreover, this condition of uncertainty is likely to become accentuated in the near future because of ongoing climate change (CC); this work sought to evaluate the impact of this change on the economics and the water management of the Nurra area . The model used for this analysis could be improved by considering the impact of additional aspects of CC (e.g., temperature, evapotranspiration and atmospheric $CO_2$) on crop cycles, and by considering the interactions among irrigation practices, network losses and groundwater recharging, which affect the water balance of the entire watershed.

Indeed, a reduction in the amount of water applied to crops does not necessarily correspond to increased water conservation, as farmers may respond to increased uncertainty regarding water availability by using improved irrigation technologies. These technologies generally enable reduced water application for a given level of crop consumption, or an increase in the area under irrigation for a given quantity of water applied. Neither outcome is a real saving of water; indeed, the latter would result in increased water consumption at the watershed level and less water availability downstream. However, in the present study area there exists little scope for improving the available irrigation technology; instead, farmers must consider making changes to cropping patterns.

Therefore, even considering the limitations of the model, the results indicate an advantage in adopting ABMs rather than VPM. The ABMs protect the groundwater resource and are consistent with the goal of setting prices that encourage farmers to use water efficiently, with the purpose of protecting the environmental quality of the resource.

## 8. Acknowledgements

Farm typologies and production technologies that characterize the agricultural sector of the area were reconstructed as part of the MONIDRI Research Project, run by INEA (National Institute for Agricultural Economics), Italy. The analysis presented in this work, however, was developed with resources of the research projects "AGROSCENARI", funded by the Italian Ministry of Agricultural, Food and Forestry Policies (MiPAAF), and "PRIN - Il recepimento della Direttiva Quadro sulle Acque (60/2000) in agricoltura" funded by the Italian Ministry of University and Scientific Research (MIUR).

## 9. References

Albiac, J. & Dinar, A. (Eds.). (2009). *The management of water quality and irrigation technologies*, Earthscan London, ISBN 978-1-84407-670-3, Sterling VA, USA

Apland, J. & Hauer, G. (1993). Discrete Stochastic Programming: Concepts, Example and a Review of Empirical Application, *University of Minnesota, Department of Applied Economics, Staff Paper*, P93–P21

Bartolini, F., Bazzani, G.M., Gallerani, V., Raggi, M. & Viaggi, D. (2007). The impact of water and agriculture policy scenarios on irrigated farming systems in Italy: An analysis based on farm level multi-attribute linear programming models. *Agricultural Systems*, 93: 90-114

Bazzani, G., Di Pasquale S., Gallerani V. & Viaggi D. (2005). Water framework directive: exploring policy design issues for irrigated systems in Italy. *Water Policy*, 7:413-28

Beguería, S., López-Moreno, J.I., Lorente, A., Seeger, M. & García-Ruiz, J.M. (2003). Assessing the effects of climate oscillations and land-use changes on streamflow in the central Spanish Pyrenees. *Ambio*, 32: 283–286

Berbel, J., Calatrava, J. & Garrido, A. (2007), Water pricing and irrigation: a review of the European experience, In: *Irrigation Water pricing Policy: The Gap Between Theory and Practice*, Molle, F. & Berkoff, J. (Eds), pp. 295-327, CABI, Wallingford UK

Brooke, A., Kendrick, D.A. & Meeraus, A. (1996). *GAMS, a user guide*. GAMS Development Corporation, Washington, D.C.

Calatrava, J. & Garrido, A. (2005). Modelling water markets under uncertain water supply. *European Review of Agricultural Economics*, 32 (2): 119–142

Cocks, K.D. (1968). Discrete Stochastic Programming. Management Science. *Theory Series*, 15, 72-79

Cornish, G., Bosworth, B., Perry, C. & Burke, J. (2004). *Water charging in irrigated agriculture. An analysis of international experience*, FAO, ISBN 92-5-105211-5, Rome

Dinar, A., Knapp, K. C. & Letey, J. (1989). Irrigation water pricing policies to reduce and finance subsurface drainage disposal. *Agricultural Water Management*, 16: 155-171

Dono, G., Marongiu, S., Severini, S., Sistu, G. & Strazzera, E. (2008). Studio sulla gestione sostenibile delle risorse idriche: analisi dei modelli di consumo per usi irrigui e civili, Collana Desertificazione, ENEA, Rome

Dono, G., Giraldo, L. & Severini, S. (2010). Pricing Irrigation Water Under Alternative Charging Methods: Possible Shortcomings of a Volumetric Approach. *Agricultural Water Management*, 97 (11): 1795-1805. DOI: 10.1016/j.agwat.2010.06.013.

Dono, G. & Mazzapicchio, G. (2010). Uncertain water supply in an irrigated Mediterranean area: An analysis of the possible economic impact of climate change on the farm sector, *Agricultural Systems*, 103 (6): 361-370, doi:10.1016/j.agsy.2010.03.005

Elbakidze, L. (2006). Potential economic impacts of changes in water availability on agriculture in the Truckee and Carson River basins, *Journal of the American Water Resources Association*, 42 (4): 841-849

European Union. (2000). Council Directive of 23 October 2002. Establishing a framework for community action in the field of water policy (2000/60/EC), *Official Journal of the European Communities*, (22 December) L327

Finger, J.M. & Kreinin, M.E. (1979). A measure of export similarity and its possible uses. *The Economic Journal*, 89: 905-12

García-Ruiz, J.M., Regüés, D., Alvera, B., Lana-Renault, N., Serrano-Muela, P., Nadal-Romero, E., Navas, A., Latron, J., Martí-Bono, C. & Arnáez, J. (2008). Flood generation and sediment transport in experimental catchments affected by land use changes in the central Pyrenees. *Journal of Hydrology*, 356: 245–260

Garrido A. & Calatrava J. (2010). *Agricultural Water Pricing: EU and Mexico*, OECD, http://dx.doi.org/10.1787/787000520088

Hardaker, J.B., Huirne, R.B.M., Anderson, J.R. & Lien, G. (Eds.), (2004). *Coping with Risk in Agriculture, 2nd ed.*, CABI, ISBN 0-85199-831-3, Wallingford UK

Iglesias, E., Garrido, A. &Gomez-Ramos, A. (2007). Economic drought management index to evaluate water institutions' performance under uncertainty. *The Australian Journal of Agricultural and Resource Economics*, 51: 17–38

IPCC. (2007). *Historical Overview of Climate Change Science*, http://www.ipcc.ch/publications_and_data/ar4/wg1/en/ch1.html

Jacquet, F. & Pluvinage, J. (1997). Climatic uncertainty and farm policy: a Discrete Stochastic Programming model for cereal-livestock farms in Algeria. *Agricultural Systems*, 53: 387–407

Levine, D.M., Krehbiel, T.C. & Berenson, M.L. (2000). *Business Statistics: A First Course, 2nd ed.*, Prentice-Hall, Upper Saddle River NJ, USA

López-Moreno, J.I., Beguería, S., Vicente-Serrano, S.M. & García-Ruiz, J.M. (2007). The influence of the nao on water resources in central Iberia: precipitation, streamflow anomalies and reservoir management strategies. *Water Resources Research* 43: W09411, doi: 10.1029/2007WR005864

Meehl, G.A., Stocker, T.F., Collins, W.D., Friedlingstein, P., Gaye, A.T., Gregory, J.M., Kitoh, A., Knutti, R., Murphy, J.M., Noda, A., Raper, S.C.B., Watterson, I.G., Weaver, A.J. & Zhao, Z.C. (2007). Global climate projections, In: *Climate Change 2007: The Physical Science Basis, Contribution of Working Group I to the Fourth Assessment Report of The IPCC*, Solomon, S., Qin, D., Manning, M., Chen, Z., Marquis, M., Averyt, K.B., Tignor, M. & Miller, H.L. (Eds), Cambridge University Press, Cambridge, New York, USA

Molle, F. & Berkoff, J. (Eds). (2007). *Irrigation Water pricing Policy: The Gap Between Theory and Practice*, CABI, Wallingford UK

Niu, X., Easterling, W., Hays, C., Jacobs, A. & Mearns, L. (2009). Reliability and input-data indiced uncertainty of EPIC model to estimate climate change impact on sorghum yields in the U.S. Great Plains. *Agriculture, Ecosystem and Environment*, 129 (1-3): 268-276

Piccolo, D. (2000). *Statistica*, Il Mulino, ISBN 8815075968

Rae, A.N. (1971a). Stochastic programming, utility, and sequential decision problems in farm management. *American Journal of Agricultural Economics*, 53: 448–460

Rae, A.N. (1971b). An empirical application and evaluation of Discrete Stochastic Programming in farm management. American Journal of Agricultural Economics, 53: 625–638

Raisanen, J. (2007). How reliable are climate models?, *Tellus Series A: Dynamic Meteorology and Oceanography*, 59: 2–29

Riesgo, L. & J. Gómez-Limón. (2006). Multi-criteria policy scenario analysis for public regulation of irrigated agriculture. *Agricultural Systems*, 91: 1-28

Semaan, J., Flichman, G., Scardigno, A. & Steduto, P. (2007). Analysis of nitrate pollution control policies in the irrigated agriculture of Apulia Region (Southern Italy): A bio-economic modelling approach. *Agricultural Systems*, 94: 357-67

Solomon, S., Qin, D., Manning, M., Chen, Z., Marquis, M., Averyt, K.B., Tignor, M. & Miller, H.L. (Eds) (2007). *Climate Change 2007 – The Physical Science Basis*. Cambridge University Press, Cambridge NY, USA

Spiegel, M.R. (1973). *Statistica*. Collana Schaum Teoria e Problemi. ETAS Libri

Tsur, Y., Roe, T., Rachid, D. & Dinar, A. (2004). *Pricing Irrigation Water. Principles and Cases from Developing Countries*, RFF, Washington, D.C., USA

Williams, J.R., Jones, C.A., Kiniry, J.R. & Spanel, D.A. (1989). The EPIC crop growth model. *Trans ASAE*, 32:497–511

# 2

# Irrigation Development: A Food Security and Household Income Perspective

Kenneth Nhundu and Abbyssinia Mushunje
*University of Fort Hare*
*South Africa*

## 1. Introduction

Rukuni, *et al*, (2006) posit that irrigation development represents the most important interface between water and land resources. Barau *et, al* (1999) stress greater emphasis on irrigation development as a means of increasing food and raw material production as well as promoting rural development. Similarly, (Hussain, *et, al*, undated) point out that agricultural water/irrigation has been regarded as a powerful factor for providing food security, protection against adverse drought conditions, increased prospects for employment and stable income, and greater opportunity for multiple cropping and crop diversification.

Furthermore, (Hussain *et., al*, undated) posit that access to reliable irrigation can enable farmers to adopt new technologies and intensify cultivation, leading to increased productivity, overall higher production, and greater returns from farming. This, in turn, opens up new employment opportunities, both on-farm and off-farm, and can improve income, livelihoods, and the quality of life in rural areas. Generally, access to good irrigation allows poor people to increase their production and income, and enhances opportunities to diversify their income base, reducing vulnerability caused by the seasonality of agricultural production as well as external shocks. Thus, access to good irrigation has the potential to contribute to poverty reduction and the movement of people from ill-being to well-being (Hussain *et, al*, undated).

Peacock (1995) defines food security as having adequate means of procuring one's basic food needs either by growing, manufacturing, mining or trading. Rukuni, *et, al* (1990) define food security as a situation where all individuals in a population can produce or procure enough food for an active and healthy life. Eicher & Staatz (1985) defined food security as a situation where all individuals in a population have access to a nutritionally adequate diet. The food security equation (Rukuni & Benstern, 1987) has two interrelated components: food availability and food accessibility. Food availability is whereby there is the availability of food through food production, storage or trade. Food accessibility is defined as the ability of the household to acquire food through production, purchases in the market from income earned or transfers.

For instance, Rukuni, *et, al* (1990) state that the largest number of food insecure households in Zimbabwe lives in natural regions IV and V, and accessing food through dry land production has been unsuccessful for most communal households given the prevailing agro-ecological factors for these regions. Populations have poor access to food because they generally lack the purchasing power that would otherwise enable them to purchase foodstuffs which they cannot cultivate. Furthermore, the incidence of food insecurity in the communal areas is largely caused by the agro ecological conditions beyond the farmers' control, high consumer prices for staple grain which erodes the household disposable income and the constraints they face in diversifying cropping patterns into higher valued cash crops.

The population densities in these natural regions IV and V have long exceeded the carrying capacity of the land, consequently leading to severe degradations of land resources in many areas, thus compromising on the efforts by smallholder farmers to break through the food insecurity trap. There are also high temperatures, lowest agricultural activities and highest incidences of agricultural failure due to frequent incidence of drought and low rainfall. The major limiting factor for the successful cultivation of crops in these regions is low rainfall and high incidence of drought. The low rainfall averages 600mm per annum, which is lower than the crop requirements for most food crops. Rukuni *et al* (1990) advocated for the need to integrate rural development interventions so as to do away with higher incidences of transitory and chronic food insecurity in smallholder communal farming areas.

Manzungu & van der Zaag (1996) postulate that one of the strategies to reduce the incidence of food insecurity in smallholder communal areas which was also advocated for by the aid organisations, policymakers, academics and lay people is a production technology appropriate for low rainfall environments. The technology is in the form of smallholder irrigation schemes. Development of smallholder irrigation schemes increases the potential for more production by counteracting mid-season dry spells and some periodic dry spells. This means that the household can grow crops more than once a year in low risk associated areas than under the rain fed production. Increased production ensures high food availability at the household level due to intensification of crop production. Intensified crop production ensures increased incomes; hence, household can purchase food, ensuring household access to food.

In this light, the Zimbabwe/European Union Micro-Project Programme (ZIM/EU MPP) has funded smallholder irrigation schemes since 1982 in Zimbabwe, but had not done any "in-depth" evaluation of the viability and impacts of these irrigation schemes, to find out whether they serve the purpose for which they were intended to and justify continued implementation of these schemes. The major objective of this study was to evaluate the impact of ZIM/EU funded irrigation projects on famers' income and food security level at Mopane Irrigation Scheme in Zvishavane District. The impact evaluation study was to justify or reform further support and investment in smallholder irrigation schemes. The study assessed the impacts on household food security and income level on a comparative analysis of irrigators and non-irrigators, and mainly looks at level of food security and incomes for both categories.

## 2. Literature review

### 2.1 Food security

Anderson (1988) points out that food insecurity may be chronic or transitory. Chronic food insecurity refers to extreme food insecurity when there is a continuously inadequate food caused by the inability to acquire food. Transitory food insecurity is whereby a household experiences a temporary decline in access to adequate food. Transitory food insecurity emanates as a result of instability in food prices, food production or people's income. In its worst form, it produces famine.

Jayne (1994) further identifies groups most vulnerable to chronic and transitory food insecurity and these include asset-poor rural people in rural and resettlement areas that farm but are often net purchasers of food. This group is said to lack the resources to produce enough income to buy their residual food requirements and this group includes female households and households in war-torn and environmentally disrupted areas, urban households with unemployed or more frequently underemployed family members. These groups typically have low levels of income and the landless labourers.

Rukuni, et, al (1990) argue that food security status among the households differs due to great variation in household s' resources and the ability to shift their resources into growth sectors with specific capital and climatic or infrastructure requirements. As a result, most smallholders in the semi-arid communal areas of natural region IV and V are not producing enough grain to meet the annual household demand. The existing literature suggests that the establishment of smallholder irrigation schemes has the potential of ensuring food security in the communal areas. Literature has also proposed different views regarding the possible impact of smallholder irrigation on food security in the communal lands.

Makadho (1994) states that the development of smallholder irrigation schemes dates back to 1912 and from 1912-1927 smallholders developed and managed their own irrigation schemes without government intervention. In 1928, the government took over some of the irrigation schemes when it felt that it was necessary to intervene in the development of this sector. Before independence, the majority of African smallholders in Zimbabwe were restricted to areas of poor soils and rainfall. The government therefore saw the development of irrigation schemes as a famine relief strategy.

Literature also suggests that earlier, the smallholder irrigation schemes had the assurance of food security at household level for smallholder communal farmers. The irrigation schemes did not only meet the intended objectives of increased food security, but also benefited the surrounding communities, who were not in the irrigation schemes. In concurrence, Rukuni (1984) reported that the areas that surrounded the schemes tended to provide a ready market for the food crops. The study by Rukuni (1984) showed that maize, beans, and vegetables had the greatest demand and were most prevalent on the schemes. About 70% percent of the maize sales were done locally.

A cost benefit analysis performed by Sithole (1995) indicated that irrigation increased household food security in the marginal to poor rainfall areas. The study also revealed that irrigation did not only improve the food security position of the level of the irrigators, but also the rest of the community benefited from these schemes. Sithole (1995) also revealed that the incomes of the irrigators were higher than the incomes of the non-irrigators. As a result of the higher incomes, the irrigation participants were in a position to purchase grain to satisfy household requirements to make up for any shortfall in production, as compared to non-participants. Sithole (1995) also compared the incomes and yields of the irrigators and that of the non-irrigators. Results of the study indicated that the smallholder schemes were both financially and economically viable and the participants were able to meet both the capital and running costs of smallholder irrigation schemes.

Sithole & Testerink (1983) conducted a study in Swaziland on the cropping and food insecurity aimed at evaluating how cash cropping contributed in alleviating food insecurity in Swaziland. The results indicated that it is only with irrigation that crop production can be carried out throughout the year in Swaziland. Sithole & Testerink (1983) concluded that increased crop production can be expected to encourage the establishment of more agro-industries to process the output, thereby increasing employment opportunities and purchasing power of individuals, implying capacity to purchase grain to meet the household requirements, thus increased food security.

A study by Gittinger et al (1990) stated that many of the world's undernourished live in large river basins in Asia, where lack of irrigation, erosion, flooding, high salinity and poor drainage represent major obstacles to improved productivity. In the semi-arid regions of Asia and Africa, the inability to harness water effectively severely limits the strength of the growing season and when the rains occur, they often take a heavy toll in flooding and soil erosion. Thus crop yields, with the existing technology of irrigation efficiency, can be doubled and increases through better control of allocation of water.

A study by Webb (1991) in a village of Chakunda in Gambia revealed that introduction of smallholder irrigation schemes increases food consumption. Webb (1991) listed the following benefits realised by participation in irrigation schemes:

- There is increased income that was translated into a boom in expenditure, investment, construction and trade.
- Backward and forward linkages resulting from traders coming to purchase irrigation produce, in this case, rice and sell cloth, jewellery and other consumables.
- Smallholder irrigation can be a worthwhile investment in the development of marginal areas of the world, coupled with the provision of irrigation facilities to communal area farmers, thus increasing yields and ensuring food security and increasing the purchasing power of the beneficiaries due to increased incomes.

## 2.2 Irrigation income

An income analysis for Mzinyathini scheme, carried out by Sithole (1995), revealed that the savings per hectare per month per household was Z$931.22 in drought relief. The income analysis for different groups, the project irrigators and the non-irrigators, suggested that the irrigators were in a better position to afford enough grain to satisfy household requirements than non-irrigators.

Meinzen-Dick et al (1993) established that among the farmers using irrigation in the natural regions IV and V, the majority (72%) were found to be food secure and had stable incomes. The study also showed that the gross margins of irrigation schemes were significantly greater than those not using irrigation. Rukuni (1985) carried out an almost similar research study in the natural regions IV and V and he showed that investment in smallholder irrigation development can have an important effect on both rural incomes and local food supplies. The results from the study revealed that the yields achieved on smallholder schemes are higher than rainfall yields in communal areas.

## 2.3 Viability of smallholder irrigation schemes

A report by Southern African Development Community (1992), mentioned that most recent schemes will not cover the cost of development and operation, thus are uneconomic. The SADC report noted that despite the support from the government and a donor, formal irrigation has not been formal. This is in controversy with some literature that suggests that smallholder irrigation scheme in marginal rainfall areas can only survive when supported by government.

This was supported by Mupawose (1984), when he was advocating for reduced subsidies on smallholder irrigation. The study further highlighted that irrigation schemes have failed and some are under-utilised. He further indicated that poor management had led to a decline in yield per unit area and to an overall lack of viability of the project. He cited that this was due to lack of interest and lack of farming experience by the irrigation participants.

In an economic analysis study carried out by Webb (1991) on smallholder irrigation scheme in Gambia, it was revealed that the increased income from irrigation resulted with increased expenditure, construction, investment and trade. A cost benefit analysis carried out by Paraiwa (1975), showed that irrigation schemes can play an important role in developing a cash economy for rural communities by making it possible for viable cash income to become accessible in a fairly large number of individuals.

A study by Peacock (1995) argued that smallholder irrigation development is not necessary for food security. The research was conducted based on comparing the cost of constructing irrigation in the communal areas and the cost of food relief coming into the area. It was shown that the costs of developing irrigation were higher than the cost of providing drought relief. The study also concluded that the development of smallholder irrigation for the purpose of food security was not economically viable.

## 2.4 Success stories of irrigation development

FAO (1997a) in a brief general overview of the smallholder irrigation sub-sector in Zimbabwe concluded that smallholder irrigation has brought success stories to farmers. The following observations were made; smallholder farmers are now able to grow high value crops both for the local and export markets, thus effectively participating in the mainstream economy, in areas of very low rainfall, as in Natural Regions IV and V, farmers enjoy the human dignity of producing their own food instead of depending on food handouts, irrigation development has made it possible for other rural infrastructure to be developed in areas which could otherwise have remained without roads, telephones, schools and clinics, smallholder irrigators have developed a commercial mentality and crop yields and farmer incomes have gone up manifold.

Similar inferences were also highlighted in a study of an irrigation scheme in the village of Chakunda in the Gambia; Webb (1991) gave the following as some of the benefits of irrigation:

- Increased income that was translated into increased expenditure, investment, construction and trade.
- Backward and forward linkages: traders were reportedly coming to purchase irrigation produce (rice) and in turn sell cloth, jewellery and other consumer items.

• Increased material wealth. At the village level, this was in the form of construction of a large mosque built through farmers' donations and an improvement of the village clinic. At household level, increased wealth could be seen in 55 houses built in the village, fourteen with corrugated metal roofing.

## 2.5 Challenges and constraints

Rukuni *et al* (2006) state that a number of problems have befallen irrigation schemes that are managed by central government departments, such as poor marketing arrangements, limited access to water, inability to meet operational costs due to poor fee structures and the lack of a sense of ownership, financial viability and poor governance. Some of these problems have necessitated government transferring responsibility to farmers, who have continued to mismanage these systems, hence their dilapidation. Poor maintenance and lack of effective control over irrigation practices have resulted in the collapse of many irrigation systems.

The FAO (1997) report identified a number of constraints, which hampered smallholder irrigation development in Zimbabwe. Some these include high cost of capital investment in irrigation works considering that communal farmers are resource poor, lack of reasonably priced appropriate irrigation technology for the smallholders, shortage of human resources at both technician and farmer levels, lack of decentralized irrigation service companies to give back-up service in rural areas, poor resource base of farmers, fragmented and small size of land holdings, unsecured or lack of land titles and high interest rates.

Further to the above constraints, Gyasi *et al* (2006) state that in many countries, institutional weaknesses and performance inefficiencies of public irrigation agencies have led to high costs of development and operation of irrigation schemes. Poor maintenance and lack of effective control over irrigation practices have resulted in the collapse of many irrigation systems. The study by Gyasi *et al* (2006) concluded that collective action for the maintenance of community irrigation schemes is more likely to be problematic when the user group size is large and ethnically heterogeneous, and where the scheme is shared by several communities. Use of labour intensive techniques in the rehabilitation of irrigation schemes promotes a sense of ownership and moral responsibility that help ensure sustainability. A high quality of rehabilitation works and regular training activities also contribute to successful irrigation management by communities.

## 3. Study area and methodology

### 3.1 Study area

It is estimated that at least 60% of Zimbabwe's communal farmers live in natural regions IV and V, where food insecurity is greatest (Rukuni, 2006). These areas are not suited to intensive farming systems. The research site was selected in natural region IV, an area with relatively less rainfall of less than 500mm and poor soils. This makes vast track of land unsuitable for cash cropping. The research was based on a case study of Mopane Irrigation Scheme, located in Runde area in Zvishavane, Midlands Province. The scheme has been functional since the year 2000 and the main crops cultivated are cash crops; wheat, maize, tomatoes and onions.

## 3.2 Sampling methods

Primary data was used as a main source of inference, while secondary data was used as a backup to the primary data. Stratified sampling was used in which the data available was divided into two strata; irrigators and non-irrigators. From each stratum, random sampling was done to obtain thirty irrigators and thirty non-irrigators. Data collection was done through structured surveys using a full administered questionnaire. The questionnaire captured data on household characteristics, asset endowment, livestock endowment, gross margin performance, agronomic practices, off-farm income, yield of grain crop. The data was entered into the Statistical Package for Social Scientists (SPSS) for further analysis.

## 3.3 Analytical frameworks

### 3.3.1 Regression analysis

A regression model was used in the regression analysis to examine the factors that affect productivity; hence food security. The project assumed the following regression model:

$$Y = \alpha_0 + \alpha_1 X_1 + \alpha_2 X_2 + \alpha_3 X_3 + \alpha_4 X_4 + \alpha_5 X_5 + U_i \tag{1}$$

Y      = Food Security
$\alpha_0, \alpha_1 - \alpha_5$ are model parameters
$X_1$      = Asset endowment
$X_2$      = Household size
$X_3$      = Off-farm income
$X_4$      = Area under cultivation
$X_5$      = Draught power ownership
$U_i$      = Random error term

The expected results from this regression model were as follows:

- Household asset endowment positively impacts food security.
- An increase in household size increases food security.
- Off-farm income has a positive impact on food security.
- Area under cultivation positively related.
- Draught power ownership enhances food security.

### 3.3.2 Gross margin analysis

Gross margin analysis was the major tool, which was used in the analysis to compare the returns between the irrigators and the non-irrigators and assess the benefits of irrigation. The study looked at the agricultural performance of both the irrigators and non-irrigators at Mopane irrigation scheme. To determine any changes in the production or productivity levels and gross incomes, a comparative analysis of inter-farm was vital. Inter-farm comparative analysis compares the irrigators and non-irrigators who are located in the same geological area.

The research study therefore used a gross margin per ha analysis as an indication of plot level performance, that is, how well farmers did on their land with the resources that were available to them. According to Johnson (1991), gross margin analysis is useful for production cycles of less than a year as this enables costs and returns to be directly linked to enterprise. Gross margin is the difference between the total sales and the variable costs.

$$\text{Gross Margin} \quad = \quad \text{Total Sales (Gross Income) - Variable Costs} \qquad (2)$$

$$\text{Where: Gross Income} \quad = \quad \text{Total Volume of Output (Q)} \times \text{Price (P)} \qquad (3)$$

and Variable Costs include the costs such as fertilizer, seed, crop chemicals, marketing costs, transport costs, machinery operational, labour costs, etc that would have been incurred in the production process until the produce has reached the market.

### 3.3.3 Farm income analysis

The crop incomes for the irrigators and the non-irrigators were derived through the use of gross margin analysis. Although the gross margin has two components that are income from sales and value of crops retained, crop output was evaluated using nominal prices. Individual household crop gross margin budgets were computed for both dry land and irrigated crops in the case of irrigators and only for dry-land for the non-irrigators. Since Mopane scheme is operated as a cooperative, only one whole farm budget was considered and then number of irrigators divided the profit to get the per income. The non-farm incomes were also compared. The main thrust behind this is to test the hypothesis that incomes of the irrigators in the project are greater than that of the non-irrigators. After computing the household gross margins, the first impressions were based on comparing the mean gross margins for the irrigators versus that of the non-irrigators.

### 3.3.4 Descriptive statistics

These were used to describe the differences between irrigation and non-irrigation households. Simple statistics like mean was employed to analyse data and yield, demographic characteristics, acreage and food availability. Also, socio-economic analysis like household size, ages, education, assets and other resources that can help in comparing the two sets of household were made use of.

## 4. Results and discussion

### 4.1 Demographic and endowment characteristics

It is vital to describe and compare household characteristics of sample households for primarily informing explanations for behavioural variability between irrigators and non-irrigators. Characteristics such as age, marital status, sex structure, employment, agricultural equipment endowment, livestock ownership, land ownership and ownership of other assets were considered important. This is because the asset base and household demographic structure of the household has implications on flexibility and capabilities with respect to crop production and consumption.

### 4.1.1 Demographic structure of households

Consideration of household demographic features offers one of the platforms on which to compare and explain behavioural variations relevant to this study.

| Variable | Irrigators (Sample Mean) | Non-Irrigators (Sample Mean) |
|---|---|---|
| Household size | 9.80 | 6.48 |
| Males | 3.44 | 2.28 |
| Females | 5.64 | 4.20 |
| Household head's age | 47 | 42 |
| Total number of children | 7.37 | 4.30 |
| Children >15 years | 4.99 | 3.02 |
| Children <15 years | 2.48 | 1.28 |
| Total no. of adults | 4.03 | 2.10 |

*Source: Survey data*

Table 1. Household Demographic Analysis

The results in Table 1 indicate that the average household size of irrigators is 9.80, higher than that of non-irrigators, with an average of 6.48 household members. There were more adults in the irrigator category with an average of 4.03 against non-irrigators' 2.10 adults. The irrigators' average household age is 47years 5years higher than that of non-irrigators (42). The irrigators have, on average more children than non-irrigators, 7.37 children per household as compared to 4.30 children for non-irrigators. This would suggest that irrigators might, on average, be more mature than the non-irrigators, who tend to be younger households on average.

Thus, the motive behind the irrigators participating in the irrigation scheme is to feed their larger household size. The larger household size may be giving the irrigators a comparative advantage, which is reflected in increasing returns to scale and decreasing average costs. For example, irrigators tend to have more labour in activities such as land preparation, where there is a great deal of labour needed, and also division of labour which increases the economies of scale.

### 4.1.2 Household land ownership

The quantity of land available per household is one of the most important constraints to production for communal farmers. Therefore, it is vital and valid to base comparison of irrigators and non-irrigators on the availability of arable land. This information is also important in that it will help in realising whether any disparities in household incomes may be accounted for by the rise in dry land holding.

| Category | Average Size of Arable Dry Land | Average Size of Irrigable Land |
|---|---|---|
| Irrigators | 2.26 ha | 0.45 ha |
| Non-irrigators | 2.09 ha | |

*Source: survey data*

Table 2. Average cropping land area

The results in Table 2 show that irrigators have more dry-land (2.26ha) on average, compared to the non-irrigators who have 2.09 ha. Under this scenario, *ceteris paribus,* irrigators are expected to have more output compared to non-irrigators. The fact that irrigators have more dry land can be attributed to the fact that they might have acquired pieces of land long before the non-irrigators, who later acquired smaller pieces of land later on. In addition to dry land, irrigators have 9ha of land, which converts to about 0.45ha per household. The irrigators do work as group and the production resources are pooled together for production and the whole produce is shared and marketed as a group.

### 4.1.3 Livestock ownership

Livestock form an important component of household food security in the communal areas. Significant differences in livestock ownership may reasonably explain differences in food security, income and agricultural technical performance between irrigators and non-irrigators as they contribute to household food availability through production, as a production asset and through household food accessibility and through income generation.

| Livestock | Irrigators | | Non-Irrigators | |
|---|---|---|---|---|
| | Sample Mean | % Owners | Sample Mean | % Owners |
| Cattle | 6.04 | 62.8 | 4.80 | 53.2 |
| Goats | 12.84 | 90.3 | 6.20 | 64.2 |
| Donkeys | 3.89 | 68.1 | 1.10 | 44.8 |
| Sheep | 0.94 | 42.7 | 0.23 | 21.8 |
| Chickens | 14.29 | 97.8 | 8.26 | 84.3 |
| Draught animals | 7.43 | 78.4 | 3.45 | 61.9 |

*Source: Survey data*

Table 3. Livestock ownership

The results in Table 3 show that irrigators have more livestock compared to the non-irrigators. Irrigators own an average of 6.04 cattle against 4.80 cattle for non-irrigators with percentage ownership of 62.8% and 53.2% respectively. Irrigators also have a higher number donkey per sample household of 3.89 compared to non-irrigators who have 1.10 donkeys. Better possession of draught animals would give the irrigators a comparative advantage in timeliness of tillage activities. Thus irrigators technically perform better than the non-irrigators, thus making the irrigators less vulnerable to poverty than the non-irrigators.

### 4.1.4 Ownership of agricultural equipment

Ownership of agricultural implements by households influences timeliness of cultivation and therefore yields. Implements can also be hired out to earn income for the households.

The results in Table 4 indicate that irrigators are better endowed with agricultural implements than non-irrigators. This implies that irrigators are wealthier than non-irrigators. However the most important tools on the farm are the plough and the hoe. Farmers often can do without such implements as scotch carts, harrows, cultivators and wheelbarrows. Since irrigators have more draught animals, it is logical and unsurprising that they also have more agricultural implements lime cultivators and scotch carts. This gives irrigators a comparative advantage in crop production in form of more timeliness in land preparation and other tillage practices. More often, the plough is used in place of a cultivator, which explains the very low number of cultivators in the two samples.

| Type of implement | Irrigators | | Non-irrigators | |
|---|---|---|---|---|
| | Sample Mean | % Owners | Sample Mean | % owners |
| Plough | 1.46 | 94.7 | 0.96 | 76.3 |
| Hoe | 6.12 | 100 | 4.31 | 100 |
| Wheelbarrow | 2.87 | 78.5 | 1.07 | 66.7 |
| Scotch cart | 0.15 | 69.7 | 0.09 | 44.0 |
| Harrow | 0.12 | 23.5 | 0.06 | 16.7 |
| Cultivator | 0.23 | 12.6 | 0.11 | 11.2 |

*Source: Survey data*

Table 4. Agricultural equipment endowment

### 4.1.5 Household housing

Two types of housing structures are dealt with in this study and these are traditional and modern houses. A traditional house is taken to be a structure, which is usually round with walls, made from mud poles or farm bricks and thatched with grass, and normally one roomed. A modern house is taken to be a rectangular structure made from farm bricks or cement bricks, zinc or asbestos roofed and constitute one or more rooms.

| Structure | Irrigators | | Non-irrigators | |
|---|---|---|---|---|
| | Sample Mean | % Owners | Sample Mean | % Owners |
| Traditional houses | 2.28 | 100 | 1.97 | 94..3 |
| Modern houses | 1.20 | 78.4 | 1.48 | 88.7 |

*Source: Survey data*

Table 5. Average number of types of housing structures of households

The results in Table 5 indicate that all irrigating households had at least one traditional house. However, non-irrigators have on average more modern houses as compared to irrigators. Also, more non-irrigators have modern houses than irrigators. The difference in modern housing may be due to the fact that since more non-irrigator household heads stay outside the village working mostly in towns or near towns, they might be bringing home the types of houses they see in towns.

### 4.1.6 Place of residence of household head

The place of residence of household head often indicates the opportunity cost of being in the village than anywhere else. In this case, the number of heads staying in the village may explain incentives attached to remaining in the village.

| Place of residence | Irrigators % | Non-irrigators % |
|---|---|---|
| Village | 56.7 | 43.3 |
| Town | 21.5 | 47.4 |
| Other | 12.8 | 9.3 |
| Total | 100 | 100 |

*Source: Survey data*

Table 6. Place of residence of household head

The results in Table 6 indicate that 47.4% of non-irrigators household heads stay away from the village, or employed somewhere outside the village than the non-irrigators who only constitute 21.5% who are in towns. This can be attributed to the fact that some non-irrigators get engaged in employment as mine workers at Shabanie Mine and other surrounding mines in Zvishavane. The higher opportunity cost associated with leaving the village and the irrigation scheme is higher than that of staying in the village, thus the irrigators are left with no other incentive other than that of staying in the village.

### 4.1.7 Household off-farm employment

Employment is defined as the number of able bodied people who are willing to work and can find a job. Table 4.7 below shows the employment status of household members.

| Employment status | Irrigators | Non-Irrigators |
|---|---|---|
| No. employed off-farm | 0.63 | 1.49 |
| % with no member in regular employment (locally or elsewhere) | 59.4 | 30 |
| % with at least one member in regular employment | 40.6 | 70 |

*Source: survey data*

Table 7. Employment status: irrigators and non-irrigators

Table 7 shows that on average, 1.49 of non-irrigators are employed off-farm as compared to 0.63 for irrigators. Off-farm employment generally indicate access to off-farm income particularly remittances. Again, 70% of the non-irrigators had at least one member in regular employment, as opposed to 40.6% of irrigators. This can be attributed to the fact that, as seen in the analysis above, more non-irrigators are employed in Zvishavane and other surrounding areas, while the irrigators see that it is more profitable to stay at the schemes, the reason why they constitute only 40.6% in regular employment.

## 5. Agricultural productivity

This subsection compares the technical performance and farm incomes to test the hypothesis that irrigators are better agriculturalists and earn more income than non-irrigators using the Gross Margin Analysis.

## 5.1 Land productivity

On average irrigators have more dry land an average of 2.26ha, 0.17ha higher than non-irrigators'. It is therefore expected that irrigators have more output than non-irrigators. The difference in land allocation may be explained by the efforts of irrigators seeking to meet the grain requirements of their larger households. Millet was more popular with irrigators for the purpose of beer brewing which was not so popular with non-irrigating younger women. Most land was devoted to sorghum among non-irrigators, which illustrates the lack of rainfall and risk of crop failure inherent in the Natural Region IV where Mopane scheme lies.

## 5.2 Dry-land production

The main source of livelihood for the farmers in Mopane area is the sale of crops. The incomes are represented in the form of gross margins, which are the incomes remaining after deducting the variable costs from the whole farm gross income.

$$\text{Gross Margin} = \text{Gross Income} - \text{Variable Costs} \qquad (4)$$

| Household Production Parameter | Price (US$/t) | IRRIGATORS | | | NON-IRRIGATORS | | |
|---|---|---|---|---|---|---|---|
| | | Ave Area (Ha) | Ave Yield (Ton/ha) | GI/Crop (US$/ha) | Ave Area (Ha) | Ave Yield (Ton/ha) | GI/Crop (US$/ha) |
| Maize (ton) | 109.10 | 0.64 | 3.500 | 381.85 | 0.59 | 3.230 | 352.39 |
| Sorghum (ton) | 563.64 | 0.77 | 0.376 | 211.93 | 0.83 | 0.418 | 235.60 |
| G/nuts (ton) | 181.82 | 0.43 | 0.466 | 84.73 | 0.36 | 0.353 | 64.18 |
| Millet (ton) | 256.97 | 0.42 | 0.351 | 90.20 | 0.31 | 0.311 | 79.92 |
| Total Av. Area (ha) | | 2.26 | | | 2.09 | | |
| Total GI (US$) | | | | 768.70 | | | 732.09 |
| GI/Ha (US$) | | | | 340.13 | | | 350.28 |
| GI/Household (US$) | | | | 11.34 | | | 10.80 |

*Source: survey data*

Table 8. Gross incomes: irrigators and non-irrigators

Maize is the most important cereal crop grown in Zimbabwe. At Mopane irrigation scheme, the crop ranks first in number of producers. As observed in the table above, there is a high yield in maize for irrigators, an average of 3.50 ton/ha, as compared to an average of 3.23 ton/ha for non-irrigators. This might be due to the fact that the irrigators, as seen in the former empirical comparative analysis, are better asset endowed than the non-irrigators, thus they perform technically better in dry land production.

However, there is a low yield of sorghum for the irrigators of 0,376 ton/ha, against 0,418 ton/ha for the non-irrigators. The irrigators grossed an average income of US$768.70 against

US$732.09 for non-irrigators from sorghum. Sorghum has better tolerance to dry conditions than maize, so non-irrigators generally devote more area to it, as a hedging strategy against food shortages.

Groundnuts yield is high within the irrigators, an average of 0,466 ton/ha compared to 0,353 ton/ha realised by the non-irrigators. This can be attributed to the fact that irrigators devote more land to its production than non-irrigators do. The difference in hectarage devoted to the crop may be explained by several factors, which include household size, total arable dry land and labour availability among others. As seen from the empirical analysis, irrigators had a comparative advantage in all of the factors above.

Irrigators have higher yields for millet of 0.351ton/ha than non-irrigators' 0.311ton/ha. It was envisaged, from informal interviews, that most irrigators are interested in income from millet through beer brewing. It was mostly older women who were interested in beer brewing, which may explain why the younger non-irrigating women were less into the crop than irrigators were. Irrigators, as seen previously, allocate more land on average for millet production than non-irrigators do. The lower yields for non-irrigators can be attributed to poor timing of cultivation activities by non-irrigators.

| Crop | Total Average Costs (US$) | |
|------|----------|--------------|
|      | Irrigators | Non-Irrigators |
| Maize (US$) | 109.77 | 75.64 |
| Sorghum (US$) | 35.39 | 36.61 |
| G/nuts (US$) | 17.48 | 11.45 |
| Millet (US$) | 19.88 | 30.79 |
| Total Var. Costs (US$) | 182.82 | 154.49 |

Source: Survey data

Table 9. Average total costs: dry-land production

Comparing the cost outlays for crop production between irrigators and non-irrigators, irrigators had significantly higher total variable costs of US$182.82 than non-irrigators' US$154.49, as shown in Table 9. It is believed that as a result of significantly higher use of variable inputs, compounded by more access to draught power and agricultural implements, irrigators had significantly higher output per ha than non-irrigators. This explains why irrigators seem to have a higher average gross margin than of non-irrigators as shown in the table 10 below.

| Parameter | Irrigators | Non-irrigators |
|-----------|-----------|----------------|
| Gross Income (US$) | 768.70 | 732.09 |
| Total Variable Costs (US$) | 182.82 | 154.49 |
| Gross Margin (US$) | 585.88 | 29,223.36 |
| Average Gross Margin (US$) | 19.53 | 19.25 |

Source: Survey data

Table 10. Gross margin analysis: dry-land production

## 5.3 Irrigation productivity

Mopane irrigation scheme produces crops during winter and summer. Total area for cropping amounts to 9ha of land. In winter, crops grown were maize, tomatoes, onions, and cabbage. Table 11 shows the hectarage allocated to each crop, average yield, price/ton, and gross income yielded, total costs in irrigation, the gross margin and the gross margin per household. Crops are grown collectively and the profits shared equally among the members.

| Crop | Total Arable (ha) | Average Yield (Ton/ha) | Price of output (US$/ton) | Gross income (US$/ha) | Total Cost (US$/ha) | Gross Margin (US$) | Gross Margin/ha (US$) | Per Gross Margin (US$) |
|------|------|------|------|------|------|------|------|------|
| Maize | 4 | 2.25 | 109.09 | 245.45 | 166.92 | 78.84 | 19.71 | 0.65 |
| Tomatoes | 2 | 4 | 181.82 | 727.27 | 295.10 | 432.17 | 216.08 | 108.04 |
| Onion | 1 | 0.86 | 96.97 | 83.39 | 16.50 | 66.89 | 66.89 | 66.89 |
| Cabbage | 2 | 2.4 | 121.21 | 290.91 | 78.90 | 212.01 | 106.00 | 56.00 |
| *Totals* | 9 | | | 1347.03 | 557.12 | 789.91 | 408.69 | 230.98 |

*Source: Survey data*

Table 11. Gross Income, Average Total Costs and Gross Margin for Irrigation

Overall, higher costs were incurred in the scheme's crop production than in dry-land production, which were US$182.82 in dry land against US$557.12 for irrigation. This can be attributed to the fact that irrigators have more income to meet these expenses and costs than the non-irrigators.

Maize is given the greatest hectarage in the irrigation scheme. An average yield of 2.25t/ha was obtained for maize. However, maize has a dry-land gross margin of US$381.85, higher than US$245.45 for irrigation. Other gross margins for other crops grown in the scheme were much higher than dry land gross margins for both irrigators and non-irrigators, indicating increased crop incomes for irrigators than non-irrigators. Main reasons for the higher yields of crops are: availability of water for irrigation during the dry season; access to water to counteract mid season dry spells, ability to extend the growing season, more agricultural implements and draught power; increased use of production inputs like fertilizer, economies of scale in resource use, for example, labour specialisation and access to technical advice from the Agricultural Research and Extension (AREX) personnel.

From table 8, it is observed that irrigators' average dry-land crop gross income per household is US$11.34, higher than non-irrigators' US$10.80. From the irrigation schemes, the gross income per participant is US$230.98 as shown in Table 11. In this respect, the irrigation scheme yields additional income for irrigators than what non-irrigators are getting from dry land farming.

## 5.4 Non-farm income

Assessing non-farm income is also important to investigate ways households supplement their income from crops. From the previous empirical analysis, it was shown that there were more non-irrigators than irrigators who stayed away from the village, employed somewhere outside the village and in Zvishavane. Though irrigators have, in terms of crop incomes outperformed non-irrigators, they might be more successful in other areas like off-farm work. As a result, there is need to evaluate and compare non-farm income of the two categories. An attempt was made to cover a number of income-earning activities in the area.

| Source of Income | Irrigators | | Non-irrigators | |
|---|---|---|---|---|
| | Mean (US$) | % of Total Income | Mean (US$) | % of Total Income |
| Remittances | 21.88 | 10.0% | 55.15 | 28.5% |
| Hiring out family labour | 29.82 | 14.7% | 26.73 | 13.8% |
| Hiring out agric implements | 24.85 | 11.4% | 5.33 | 2.8% |
| Sale of livestock | 22.30 | 10.2% | 14.55 | 7.5% |
| Building activities | 32.12 | 14.7% | 46.73 | 24.1% |
| Beer brewery | 16.36 | 7.5% | 12.42 | 6.4% |
| Cross Boarder | 30.30 | 13.9% | 17.58 | 9.1% |
| Shop business | 40.48 | 18.6% | 15.09 | 7.8% |
| **Totals** | **218.12** | **100%** | **193.58** | **100%** |

*Source: survey data*

Table 12. Other sources of household income

Table 12 above examines the other sources of income besides cropping. Remittances were vital in non-irrigators with 28.5% contribution to total income, compared to 10.0% for irrigators. This is because more members from non-irrigating households are in regular employment as previously shown in Table 7. The highest income earner to irrigators is shop business, representing a contribution of 18.6% compared to 7.8% for non-irrigators. However, building activities tend to contribute significantly to both irrigators and non irrigators, with a contribution of 14.7% and 24.1% respectively.

Irrigators have more income on average, (US$218.12) against US$193.58 for non-irrigators. This can be attributed to the fact that irrigators have more livestock, which they sell as reflected by a proportion of 10.2% for irrigators compared to 7.5% for non-irrigators, and more agricultural implements, which they hire out. The larger size of the irrigators also gives them the opportunity of hiring out family labour which also contributes to the average income for irrigators as compared to non-irrigators.

Some females, from both categories are also involved in trading activities where they go to countries like South Africa where they buy other goods for resale. This contributes significantly to both the incomes of both, though female irrigators gross more from such activities. It is also important to say that since irrigating households are bigger and older they have greater division of labour and diversified off-farm income sources. This confirms that income of irrigators is greater than that of non-irrigators since the irrigators have more income in dry land and irrigation activities as compared to the non-irrigators.

## 5.6 Regression analysis results

Applying the regression model, the econometric results are presented as in Table 13 below. The dependent variable is food security. The estimates indicate essentially in accordance with the hypothesis that the irrigators are more food secure as compared to the non-irrigators. The variables in the model that affect household food security include household size, sex of household head, off-farm income, area under cultivation and draught power ownership. Each parameter estimate measures the relationship or contribution of each variable to the food security level per household.

| Independent Variable | Parameter Estimate | T- value | Significance |
|---|---|---|---|
| Intercept (constant) | - 45.326 | - 0.429 | 0.528 |
| Household size | 88.423 | 2.914 | 0.107* |
| Household asset endowment | - 31.853 | - 1.495 | 0.163 |
| Off-farm income | 5.265 | 2.480 | 0.14* |
| Area under cultivation | 0.839 | 3.486 | 0.0485* |
| Draught power ownership | 9.202 | 2.146 | 0.058** |
| Random error term | 86.574 | | |

Source: survey data

Table 13. Regression analysis model and the estimates

$$R^2 = 0.718 \qquad \text{Adjusted } R^2 = 0.641$$

* - indicate significance at the 5% level
** - indicate significance at the 5% and 10% level

The results indicate $R^2$ is 0.718, implying a degree of 71.8% relationship among the independent variable. The adjusted $R^2$ shows that 64.1% of the variables can explain the model and the higher the adjusted $R^2$, the more significant the model. Therefore, the variables can significantly explain the model.

Household size, as can be seen Table 13 is significant at the 5% level and the positive coefficient indicates that there is a positive relationship between food security and household size. It was observed in the previous analysis that irrigators were seen to have a higher household size on average than the non-irrigators. This explains why food security increases with an increase in household size since more labour will be available to work in the irrigation and dry land plots, including hiring out labour and raise income to purchase more food. This supports the hypothesis that irrigators are more food secure and higher incomes compared to non-irrigators.

Off-farm income is also significant at the 5% significant level and the coefficient is positive. This indicates that an increase in off-farm income leads to an increase in the food security. As previously observed in the preliminary analysis of the study, the irrigators had more off-farm income than non-irrigators, thus it can be concluded that they are more food secure than the non-irrigators. This again supports the hypothesis that irrigators are more food secure than non-irrigators.

The area under cultivation is also seen to positively affect household food security. This is shown by a positive coefficient in the model. This means that as area under irrigation increases, household food security also increases. It is also, at the 5% significance level true that irrigators are more food secure compared to non-irrigators. This is because the irrigators were seen to own, on average more land than non-irrigators did, coupled with that from irrigation. This can be attributed to the fact that they can produce more per given area, thus boosting their food production for the family.

Draught power is also another variable that is seen to positively affect the level of household food security. This is significant at the 5% level and can safely support the hypothesis that irrigators are more food secure since they were seen to have more draught power on average than non-irrigators. As a result, they engage in timeliness ploughing, thus aiding in boosting output production.

## 6. Conclusions

### 6.1 Socio-economic characteristics of the household

Irrigators were found to be larger households and older than non-irrigating households. Non-irrigators had more members in regular employment than irrigators, suggesting more income to non-irrigators from remittances. Irrigators have more livestock on average than non-irrigators. On agricultural equipment, irrigators were better endowed than non-irrigators were. On housing, non-irrigators' houses were more modern as compared to those of irrigators. Finally, irrigators had more land than non-irrigators suggesting increased production of more food from dry land cropping than non-irrigators. Non-irrigators seem to be more into off-farm regular employment than irrigators.

### 6.2 Impact on food security

The study has presented some evidence to show that irrigators produce more food than non-irrigators. The output of irrigators from dry-land and irrigation is greater than non-irrigators' output from dry land production. The irrigators were also seen to have more dry-land on average, coupled with that from irrigation as compared to non-irrigators. As a result, they had more crop output compared to the non-irrigators. This ensures availability of food for them. From the gross margin analysis, it was seen that irrigators had more crop income, and coupled with non-farm income, they have more disposable income, which they can use for purchasing household food requirements which cannot be locally produced.

The irrigation scheme has also been seen as a source of food where non-irrigators would buy the produce like cabbage, tomatoes and onions. Thus, irrigators have more disposable income as compared to non-irrigators. More income implies a much better security position for irrigators giving them the opportunity to purchase more nutritious foods. As was observed, the farmers grow cabbages, onions and tomatoes and these crops do help in relieving malnutrition. Thus, the hypothesis that irrigation increases the food security level in the communal areas is therefore accepted, provided that food markets are available.

### 6.3 Impact on farm incomes

It has been shown from the study that irrigation increases the incomes of the smallholder irrigation farmers through crop incomes. This was done on a comparative analysis scenario where the gross margins from dry land for both the irrigators and non-irrigators were computed. The larger contribution of income from irrigation has evidenced that the irrigation scheme increased the incomes of irrigators substantially, and was largely responsible for the significant difference in the income levels between both categories. Higher incomes improve the standard of living; hence irrigation improved the welfare of irrigators.

The evidence supports the hypothesis that irrigators have more income as compared to non-irrigators. An analysis of other sources of income was conducted and showed a higher off-farm income for irrigators than of the non-irrigators'.

## 6.4 Technical performance

Smallholder irrigation schemes increase agricultural productivity. Irrigators were seen to perform better than non-irrigators. This is attributed to the fact that irrigators are better factor endowed, had more draught power and labour force. This means they practiced timeliness agricultural activities, thus increasing agricultural productivity. Irrigators also have better access to extension services through AREX personnel who constantly disseminate information to them, unlike non-irrigators who often meet him after a long period. Thus, we fail to reject the hypothesis that irrigators are better agricultural performers than non-irrigators.

## 7. Policy insights

Irrigation, as has been established from this study, positively impacts on the irrigators through improving household food security and income, hence standard of living for the irrigators. As a result, ZIM/EU MPP, together with the government and private sectors, should be encouraged to invest more in smallholder agriculture. Increases in the incomes realised from irrigation scheme contributes to the Gross Domestic Product, which is an aspect of economic growth. Hence, irrigation contributes to economic growth of the nation.

The irrigation scheme was seen to make a positive contribution to household food security, thus, it is a way of ensuring that people have access to adequate, nutritious food in their homes. This improves on the standards of living of the rural poor.

## 8. Recommendations

The study shows that smallholder irrigation can make a significant contribution towards poverty alleviation, increased incomes and food security. As such, ZIM/EU MPP and other donor NGOs should continue and be encouraged to support smallholder irrigation scheme investments. This should spread to all areas in the country, especially to those communal areas where rainfall is erratic. This will ensure food security, increased incomes, improved standards of living and employment creation for the rural population.

Governments, public and private institutions and non-governmental organisations are recommended to work together defining and implementing comprehensive strategies for smallholder irrigation development especially in the smallholder communal areas so as to ensure food security and employment to the rural population. There is need to formulate a comprehensive strategy to promote small-scale irrigation, including the accessibility of appropriate and affordable technology.

Such a strategy should include the following components:

- Review existing regulations and policies that influence small-scale irrigation.
- Define the role of government institutions, private sector and non-governmental organizations (NGOs) in promoting the adoption of improved irrigation technologies by small farmers. The private sector and NGOs should be encouraged to participate. However, it is recognized that government should play an active part in the identification and development of appropriate technologies and in the wider issues of rural infrastructural development so as to encourage expansion of smallholder irrigation projects.
- Encourage private investment in irrigation through provision of credit and financial incentives targeted to smallholder irrigation.
- The local rural district councils should make sure that they get in touch with NGOs, like ZIM/EU MPP and the donor community willing to take part in establishment and development of smallholder irrigation schemes, leading to self-sufficiency and food security.

## 9. References

Anderson P. P. (1988). Food Security and Structural Adjustments. World Bank, Washington, D.C.

Barau, A.D., T.K. Atala And C.I. Agbo (1999). Factors affecting efficiency of resource use under large scale farming: A case study of dadin kowa irrigation project, Bauchi State. Nig. J. Rur. Econ. Soc., 1: 1-6.

Eicher C. & Staatz J. (1985). Food Security Policy in Sub-Saharan Africa. Invited Paper Prepared For The XIXth Conference Of The International Association Of Agricultural Economists, August 25 – September 5 1985. Malanga, Spain.

Gittinger, J., Chernick, S., Hosenstein, N.R. and Saiter, K., 1990. Household food security and the role of women. In: *World Bank Discussion Paper No. 96*, the World Bank, Washington, DC.

Hussain I, Mark Giordano M, & Munir A (Undated). Agricultural Water and Poverty Linkages: Case Studies on Large and Small Systems. http://www.iwmi.cgiar.org/propoor/files/ADBProject/ResearchPapers/Agricult ural water poverty_linkages.pdf. Accessed 01 August 2010.

Jayne T. S. (1994). Market-oriented Strategies to Improve Household to Food Experience from Sub-Saharan Africa. MSU, International Development Working Paper No. 15, Department of Agricultural Economics, Michigan State University.

Johnson D. T. (1991). The Business of Farming. A Guide to Farm Business Management in the Tropics. McMillan Publishers Ltd, London.

Makadho J.M. 1994. An Analysis of Water Management Performance in Small holder Irrigation Schemes in Zimbabwe. A dissertation submitted to the Department of Soil Science and Agricultural Engineering in partial fulfillment of the requirements for the degree of Doctor of Philosophy. University of Zimbabwe.

Manzungu E & van der Zaag P (1996). The Practice of Smallholder Irrigation. Case Studies from Zimbabwe. University Of Zimbabwe Publication Office, Harare.

Meinzen-Dick R, Makombe G & Sullins M. (1993). Agro-economic Performance of Smallholder Irrigation in Zimbabwe. Paper Prepared For UZ/Agritex/IFPRI Conference On Irrigation Performance In Zimbabwe, 1-6 August 1993 At Montclair Hotel, Juliasdale.

Mupawose R.W. (1984). Irrigation in Zimbabwe. A broad overview in Africa Regioal Symposia on Smallholder Irrigation. Edited by Blackie, 1987, University of Zimbabwe, Harare.

Paraiwa, M.G. (1975). *An economic analysis and appraisal of Chilonga rrigation scheme in Matibi Tribal Trust Land.* Economics and Marketing Branch, Ministry of Agriculture. Zimbabwe

Peacock T. (1995). Financial & Economic Aspects of Smallholder Irrigation in Zimbabwe & Prospects for Future Development in Water Development for Diversification within Smallholder Farming Sector. Paper presented at Monomotapa Hotel, Harare, May 30 1995. Zimbabwe Farmers Union.

Rukuni M & Benstern R. H. (1987). Major Issues in Designing a Research Programme on Household Food Insecurity in Southern Africa. Food Security Policy Option. UZ/MSU.

Rukuni M & Eicher C. K. (1987). Food Security for Africa. University Of Zimbabwe, UZ/MSU. Food Security Project. University of Zimbabwe Publications, Harare.

Rukuni M, Eicher C.K & Blackie (Eds). (2006). *Zimbabwe's Agricultural Revolution, Revisited,* University of Zimbabwe Publications, Harare.

Rukuni, M. 1984. Cropping patterns and productivity on smallholder irrigation schemes. In: Blackie, M.J. (editor) *African regional symposium on smallholder irrigation.* University of Zimbabwe.

Rukuni, M., Mudimu, G. and Jayne, T. (eds.) 1990. *Food Security Policies in the SADCC Region.* Proceedings of the 5th Annual Conference on Food Security Research in Southern Africa, 16-18 October, University of Zimbabwe, Harare.

SADC (1992). Regional Irrigation Development Strategy. Country Report, Zimbabwe. ACK Australia Pvt Ltd, Harare.

Sithole P.N (1995). The Impact of Smallholder Irrigation on Household Food Security in Marginal Areas.

Sithole, V.M. and J. Testerink. (1983). "Irrigation and food security in Swaziland: current status and research priorities". Paper presented at the 4th Annual University of Zimbabwe/Michigan State University Conference on Food Security in Southern Africa, 31 October – 4 November, 1983. Social Science Research Unit, University of Swaziland, Swaziland.

Webb (1991). When Projects Collapse; Irrigation Failure in Gambia From a household Perspective. Reprinted From Journal Of International Development. Volume 3. No 4 July, Washington DC, USA.

www.cso.co.zw, Real Values, Nominal Values and the Price Index, Accessed 18 September 2008.

Webb P (1991). *When projects collapse: Irrigation failure in the Gambia from a household perspective.* Journal of International development Vol. 3, No. 4. July Institute, Washington D.C.

GYASI, ENGEL & FROHBERG (2006). *What determines the success of community-based institutions for irrigation management?* ZEF Policy Brief No. 5, Ghana

# 3

# Equity in Access to Irrigation Water: A Comparative Analysis of Tube-Well Irrigation System and Conjunctive Irrigation System

Anindita Sarkar

*Department of Geography, Miranda House, Delhi University, City, Delhi*
*India*

## 1. Introduction

Access to irrigation water is one of the most important factors in modern agricultural production. It offers opportunities for improving livelihoods particularly in rural areas as access to reliable good quality irrigation reduces the cost and increases the quantum of production by reducing the risks faced by the rain fed agriculture. In agricultural water distribution, equity is limited to allocation and receipt of irrigation water. Equity means fairness in creating fair access to water for all, both within and between communities and within and between regions. Since more than 60% of the irrigated area is under groundwater and is fast increasing with time, the equity in access to groundwater is of great concern. It is noteworthy to mention that in the Indian context water allocation principles refer to 'proportionate equality' and 'prior appropriation'. The former operates in the existing inequality of land ownership and the later generates inequality through uses (Pant 1984).

By the very nature of the resource, groundwater development is largely by private initiative of farmers which is conditioned by their size of land holding, savings and investment capacities. Because of this reason in the first phase of groundwater exploitation, the poor invariably got left out in the race for groundwater irrigation and decades later when they began to enter groundwater economy a set of new rules and regulations like licensing, sitting rules and groundwater zones made their entry difficult in most areas and impossible in those areas where groundwater overdraft was high (Shah 1993). With intensive groundwater exploitation, declining water tables have further reduced access to groundwater irrigation to a large number of small and marginal farmers who can neither use traditional techniques nor are able to use 'lumpy' new technology so as to pump water at an economic price. Moreover, chasing water table is beyond the reach of resource poor farmers. In such conditions they have to depend on the other well owners for groundwater irrigation. This has severe equity implications especially in a situation where farmers have little opportunity to earn their income from sources other than irrigated agriculture (Dhawan 1982). Thus in the process, the race to exploit groundwater resource is exponentially continued by the haves and the have-nots continue to bear the brunt of this negative externality (Nagraj and Chandrakanth 1997). As a consequence, there emerges widespread apprehension that, instead of reducing relative inequalities among rural incomes, groundwater irrigation development may actually have enlarged both the absolute and relative inequalities already prevalent (Shah 1987 and Shah 1993). Many micro level studies have also highlighted these serious equity implications of groundwater exploitation with falling water levels particularly in the water-starved regions (Shah 1991, Bhatia 1992, Monech 1992, Nagraj and Chandrakanth 1997). While groundwater availability can be studied from an earth science perspective but to analyse its accessibility one needs deeper understanding of groundwater economy and its underlying socio economic dynamics.

The policy design aimed to achieve food security of the country in the sixties encouraged "grain revolution" with increasing area under water intensive rice-wheat cropping pattern in the Green Revolution belt making Punjab the 'Bread basket' of the country. During this time, the modern agricultural practices of HYV technology in Punjab also ushered in the shift from canal irrigation to tube-well irrigation as it was a more reliable and flexible source of irrigation and this gave boost to enormous increase in agricultural production. In the early phase of Green Revolution, rapid diffusion of groundwater technology was thus appreciated on grounds of it being economically superior to other sources of irrigation in terms of its efficacy and productivity (Dhawan 1975). The superiority of this irrigation source continued to enhance the intensive cultivation of water intensive crops on an extensive scale not withstanding the hydro-geological thresholds of this resource. Consequently the over exploitation of groundwater inevitably questions the accessibility of this resource and rises serious concerns about the equity in its distribution.

Literature highlighting the superiority of the modern water extraction machines has been too preoccupied with highlighting the superiority from individual or private point of view which only focuses on economic justification and economic efficiency without considering the economic equity. It should be noted that economic efficiency begins to introduce a concern for equity that was missing in economic justification, in the specification that the increase in welfare of one individual should not be at the expense of another. The economic justification although assures enough benefits generated to cover all the costs but do not take into account the economic equity criterion which requires the costs to be allocated in proportion to benefits received (Abu-Zeid 2001).

In this broader context, the paper examines three aspects inequity in access to groundwater irrigation across different classes of farmers in different phases of groundwater depletion in Punjab. The study analyses the external diseconomies in groundwater utilization in terms of its accessibility to groundwater irrigation to large farmers vis-à-vis the small and marginal farmers. Firstly, it looks into the determinants of groundwater accessibility. Secondly, it empirically shows the difference in the physical and economic accessibility of groundwater resource and thirdly it evaluates the consequences of unequal access to groundwater irrigation by analysing the inequity in net returns to agriculture among agricultural communities dependent on groundwater irrigation.

Since depletion is a phenomenon, to capture the effects of groundwater depletion, in this study three villages are chosen from the same agro-climatic region with different levels of groundwater depletion. Three hundred households are interviewed from each village to collect field level data for the analysis. Table 1 gives the profile of the three study villages and figure 1 shows their locations.

Equity in Access to Irrigation Water: A Comparative Analysis of Tube-Well Irrigation System and
Conjunctive Irrigation System

51

| Name of the Village | Tohl Kalan | Gharinda | Ballab-e-Darya |
|---|---|---|---|
| Slope | Gentle | Gentle | Gentle |
| Prevalent Soil Type | Alluvial | Alluvial | Alluvial |
| Average depth of water table below | 12 meters | 18 meters | 46 meters |
| Type of irrigation | Mixed | Groundwater | Groundwater |
| Sources of Irrigation | canals – 43 %<br>tube-wells – 57 % | tube-wells – 100 % | tube-wells – 100 % |
| Cropping Intensity (%) | 204 | 217 | 178 |

Table 1. Profile of Study Areas

## 2. Determinants of groundwater accessibility

Studies have indicated that ownership and access to groundwater irrigation has almost replaced land in determining one's socio-economic and political status (Janakarajan S. 1993). In the groundwater dependant societies, the struggle for access to, and control over groundwater, shapes the course of agrarian change and development (Dubash 2002). Certain factors which govern the ownership of groundwater are central to understanding changes in access to groundwater over time. Under British common law, the basic civil law doctrine governing property ownership in most of India, groundwater rights are appurtenant to land (Singh 1992). If a person owns a piece of land, he/she can drill or dig a well and can pump out as much groundwater as he/ she is able for use on overlying lands. When land is sold the groundwater access rights pass with the land and can not legally separated from it. At present, groundwater rights are defined by the ability to chase water tables and ability to invest in changing water technology. If one can afford to deepen ones well, the water pumped out from it is theirs (Moench 1992). Groundwater accessibility is thus largely depend on a wide interplay of interconnected factors like land holding size, type and nature of ownership of tube-wells, productivity of tube-wells and density of tube-wells. The following section analyses the interplay of these dynamic factors among various size classes of farmers at different levels of groundwater depletion to understand the variability of groundwater accessibility with continuous resource depletion.

### 2.1 Land ownership and accessibility to groundwater

The distribution of land ownership and the extent of land subdivision and fragmentation affect the development and use of groundwater. Jairath (1985) argues that fragmentation of landholdings has led to underutilization of privately owned tube-wells in Punjab. Thus large farms may more beneficially utilize groundwater irrigation structures than the small ones. Moreover the higher farm productivity of large farms also facilitate the greater investments in buying and maintaining tube-well technology which is essential for continued accessibility of groundwater irrigation (Dubash 2002). Inequalities in the ownership of water extraction machines are closely related to the inequalities in land ownership and the inequalities in land and water ownership are seen to compound each other (Bhatia 1992). Thus the pattern of land ownership inevitably influences the farmers' ability to access groundwater and since availability of groundwater varies according to the levels of the existing water table, it is important to examine how different land holding categories at different levels of resource depletion differ in access to groundwater irrigation.

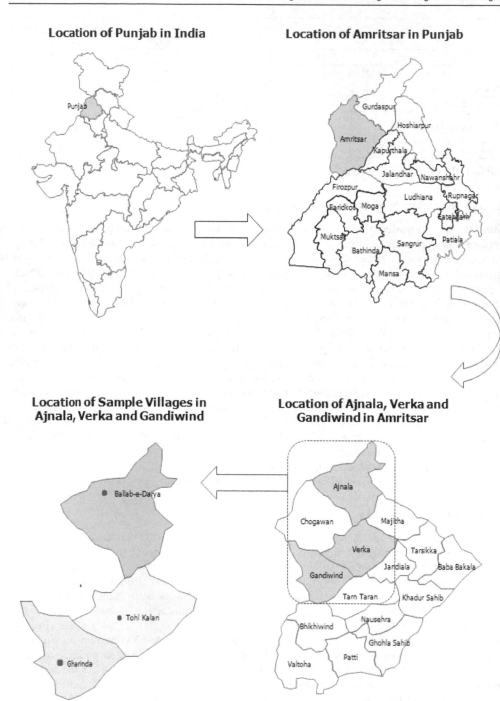

Fig. 1. Location of Study Areas

Equity in Access to Irrigation Water: A Comparative Analysis of Tube-Well Irrigation System and
Conjunctive Irrigation System

53

If we accept land as a reasonably good indicator of power in agrarian societies, then all the sample villages are societies with deep inequalities of power (Table – 2). Landownership and land operation through tenancy are linked in a way that they defy easy separation. Studies show that in Punjab "reverse tenancy" is a common phenomenon under which small and marginal farmers lease out land on cash terms to the medium and large farmers who have sufficient capital and have made investment in machinery and in water extraction machines (Siddhu 2002). A careful examination (Table–3) reveals that reverse tenancy is

| Land owned (acres) | Mixed Irrigation Village (Tohl Kalan) | Tube-well Irrigation Village (Gharinda) | Tube-well Irrigation Village with Problems of Depletion (Ballab-e-Darya) |
|---|---|---|---|
| 1 - 2 | 18 | 4 | 32 |
| 2 - 4 | 26 | 6 | 15 |
| 4 - 10 | 38 | 32 | 32 |
| more than 10 | 18 | 58 | 21 |
| Total | 100 | 100 | 100 |

Source: Questionnaire surveys in various villages from May to July, 2009

Table 2. Land Ownership by Different Classes of Farmers (Percentages)

| Land owned (acres) | % of households in each group | % of households leasing in | % of households leasing out | leased in area as % of operated area | leased out area as % of operated area |
|---|---|---|---|---|---|
| Mixed Irrigation Village (Tohl Kalan) | | | | | |
| 1 - 2 | 18 | 11 | 6 | 9 | 21 |
| 2 - 4 | 26 | 23 | 0 | 11 | 0 |
| 4 - 10 | 38 | 13 | 0 | 6 | 0 |
| more than 10 | 18 | 50 | 0 | 28 | 0 |
| Total | 100 | 22 | 1 | 18 | 2 |
| Tube-well Irrigation Village (Gharinda) | | | | | |
| 1 - 2 | 4 | 0 | 0 | 0 | 0 |
| 2 - 4 | 6 | 0 | 0 | 0 | 0 |
| 4 - 10 | 32 | 9 | 0 | 5 | 0 |
| more than 10 | 58 | 17 | 2 | 5 | 9 |
| Total | 100 | 13 | 1 | 5 | 7 |
| Tube-well Irrigation Village with Problems of Depletion (Ballab-e-Darya) | | | | | |
| 1 - 2 | 32 | 0 | 28 | 0 | 29 |
| 2 - 4 | 15 | 0 | 7 | 0 | 9 |
| 4 - 10 | 32 | 13 | 0 | 5 | 0 |
| more than 10 | 20 | 20 | 0 | 16 | 0 |
| Total | 100 | 8 | 10 | 10 | 2 |

Source: Questionnaire surveys in various villages from May to July, 2009

Table 3. Incidence of Tenancy by Landownership (percentage of land leased out to total land owned by each group)

prevalent in the sample villages and it is also more pronounced in the tube-well irrigated village of Ballab-e-Darya indicating close correspondence of this phenomenon with groundwater depletion.

Field observations reveal, in the tube well irrigated regions of Punjab, the small farmers who do not have their own source of irrigation and are also not in a position to buy water for irrigation are compelled to lease out their land to the large farmers especially in the kharif season when there is acute water scarcity on account of rice cultivation[1]. In spite of much exploitation, farmers prefer leasing out land in kharif season because it is still more profitable than rain-fed maize cultivation. The value of the land is calculated purely on the basis of availability of water supply for irrigation which in turn depends on the number of wells in that particular land, its depth and the capacity of the pump used to pump out water. It was seen that land endowed with sufficient groundwater irrigation was leased out at Rs 16,000 to Rs 20,000 per acre and land without any source of water was leased out for Rs 6,000 to Rs.8,000.

Very exploitative tenancy relations were also common in lands without any water extraction machines. In such cases the owner (mostly small or marginal farmers) pays for all the inputs like seeds, fertilizers, insecticides, labour and the produce is divided equally between the owner and the tenant. The tenant who is a large land lord only provides with the irrigation water and takes away half of the produce. Thus, ownership of groundwater determines the terms and conditions of tenancy in groundwater depleted regions in Punjab. These indicate that with groundwater depletion, water becomes the most important factor of cultivation and even its importance exceeds that of land. In such groundwater dependant societies, land has no value unless it is endowed with water extraction machines and the bargaining power is also in the hands of those who own water along with land and not only land. Thus, there is a complete shift of power relation from the hands of 'landlords' to 'waterlords'.

The control and access over groundwater offers scope for interlinkages between ownership of land and water. Such 'interlinked contracts' have been observed for land, labour and credit, and similar contractual forms in the provision of irrigation may be an additional mechanism of marginalising resource poor to groundwater access. The link between credit and groundwater has several possible implications. Usurious credit relations driven by groundwater related investment, carry the potential for a long term debt trap. They also allow a creditor to dictate production decisions especially the decisions of cropping pattern. Creditors are mostly landowners, leading to credit relations being 'interlinked' with land and water arrangements in various combinations. In the villages of Punjab, such interlinked 'land-water-credit markets' were very common especially in regions of acute depletion. Interlinkages between these three important determinants of cultivation have lead to sever consequences in accessibility to groundwater irrigation and hence a profitable agriculture. Institutional credit is not available to set up new tube-wells and land without water can not be cultivated. Farmers owning smaller assets (lands) thus often fall prey to local money lenders. As cost of inputs increase with time, credits become a necessary condition to sustain cultivation. The farmers owning small land holdings without any water extraction machine have no alternative option but to take loan from local money lenders or lease out or sell out

---

[1] Rice and maize are grown in Kharif season. But the relative profitability of growing rice is much higher than maize. This is a half yearly lease.

Equity in Access to Irrigation Water: A Comparative Analysis of Tube-Well Irrigation System and
Conjunctive Irrigation System

55

his land. Thus, the interlinking of credit, land and water leads to much greater exploitation of the less endowed farmers and in the process they lose their land and turn into agricultural labourers or construction workers in urban areas from a cultivator.

## 2.2 Ownership of wells and access to groundwater

In agrarian societies heavily reliant on irrigated agriculture, control over water is an essential complement to landownership (Dubash 2002). Available evidences in literature indicate strong positive correlation between land holding size and ownership of modern water extraction machines (Shah 1988) which is also true in all the three sample villages (Table - 4). Since the development of a well for irrigation requires substantial investments, it is largely affordable by the resource rich farmers who are also the large landlords. This implies that better access to land is associated with the better access to groundwater. Along with this, the inequality in the distribution of operational tube-wells is most pronounced in the groundwater depleted village because with receding water tables more numbers of wells of small and marginal farmers dry up as they have no capital to chase water table. Positive correspondence with landholding size and average depth of tube wells and average land irrigated per bore well reiterating the same findings (Table - 4). Thus, along with the inherent inequality of tube-well ownership influenced by the unequal distribution of land ownership, groundwater depletion further increases the skewedness in the ownership of tube-wells.

| Particulars | Marginal Farmer | Small Farmer | Medium Farmer | Large Farmer | Total |
|---|---|---|---|---|---|
| Mixed Irrigation Village (Tohl Kalan) | | | | | |
| Average no of operational tube-wells (feet) | 0.72 | 0.96 | 1.00 | 1.00 | 0.94 |
| Average Depth of tube well | 180 | 191 | 231 | 421 | 249 |
| Average land irrigated per bore well (acre) | 4.42 | 7.05 | 12.99 | 40.56 | 15.50 |
| Tube-well Irrigation Village (Gharinda) | | | | | |
| Average no of operational tube-wells (feet) | 1.00 | 1.00 | 1.06 | 1.34 | 1.22 |
| Average Depth of tube well | 120 | 185 | 210 | 217 | 209 |
| Average land irrigated per bore well (acre) | 2 | 4 | 7 | 13 | 10 |
| Tube-well Irrigation Village with Problems of Depletion (Ballab-e-Darya) | | | | | |
| Average no of operational tube-wells (feet) | 0.41 | 1.00 | 0.91 | 1.80 | 0.94 |
| Average Depth of tube well | 120 | 185 | 210 | 217 | 209 |
| Average land irrigated per bore well (acre) | 2 | 4 | 7 | 13 | 10 |

Source: Questionnaire surveys in various villages from May to July, 2009

Table 4. Tube Well Ownership and Area of Influence of Tube Wells across Farm Size Classes (Change into percentage)

Moreover, the poor farmers even after owning wells may be trapped in a regime of low well yields as not only water table is receding progressively but also many new wells are dug[2]. Because of declining water tables and increasing density of wells, it is difficult to access a new location to fix up a new well which is a necessary condition to avoid well interference and hence have a productive well. Large farmers owning large plots of land have greater opportunity to space his wells. On the contrary, the small and marginal farmers have little option to get a suitable place to dig his well as he owns a small fragment of land and very often he is a late initiator of the tube-well technology and the neighbouring plots already have deep tube-wells.

## 2.3 Nature of ownership of wells and access to groundwater

In Punjab, some of the most important factors affecting access to groundwater irrigation include whether wells are owned solely by individuals or held jointly. It is seen that the average individual ownership of tube-wells is much higher for large landowners than the marginal and small land owners (Table -5). The strong preference of individual ownership of tube-wells despite the higher costs involved reflects that individual exploitation of water even at higher costs is sufficiently productive to be economical. Individuals may also be prepared to bear higher costs because of difficulties in ensuring effective joint ownership and management of wells, and the risks depending on purchases from other tube-well owners. In conditions of continuous groundwater mining even available supplies are inadequate to meet the demand of the area served by an aquifer, these constraints become more severe (Janakarajan and Moench 2006). This fact is also reinforced by the much higher average number of sole ownership of tube-wells in the groundwater depleted village of Ballab-e-Darya than in the other two villages (Table-).

The incidences of hiring of tube-wells were not common phenomena in the villages because land and water extraction machines was considered as complementary resource and the leasing in and leasing out of land automatically resulted in the leasing in and leasing out of the tube-well in the respective land. Hiring of tube-wells also does not show any correspondence with land holding size. With groundwater depletion the farmers do not want to hire wells as disputes arise as to which party will deepen the well and repair the pump which becomes a hurdle for timely irrigation. The farmers, thus, prefer to lease out the entire land and tube-well to have complete control and responsibility of the tube-well. Due to these impediments of groundwater accessibility through hired tube-wells, hiring has become redundant in the villages of Punjab.

Since tube wells are indivisive, with successive generation number of land holdings increase and the numbers of shareholders consequently increase in a family owned well. Sometimes even the partners (subsequently the heirs of the partners) of the old water extraction technology like *hult*[3] continue to jointly irrigate and own wells. In many cases especially for newly owned joint wells, either the brothers and cousins or neighbouring farmers owning small fragments of (contiguous) land contribute jointly to install submersible pumps. Joint wells are commonly operated by installing a single pump set

---

[2] With many wells, the density of tube-wells increases lowering the yield of the neighbouring wells.
[3] *Hult* was a traditional water extraction machine and it needed lot of labour (both animal and human) to irrigate land. As it was labour intensive families jointly owned and operated *hults*.

Equity in Access to Irrigation Water: A Comparative Analysis of Tube-Well Irrigation System and
Conjunctive Irrigation System

57

and running the motor in rotation between shareholders for a fixed number of hours. It helps them to share the cost and also fully utilize the chunk of economic investment for (jointly) irrigating the combined portion of land. With the incresing number of joint ownership of wells, the dilemma and uncertainties associated with management of jointly owned wells create varied nature conflicts within communities and families which is important to analyse as it revolves round several issues of equity to accessibility of irrigation water among the shareholders.

| Land Holding Category | Solely Owned Tube-Wells | | Hired Tube-Wells | | Jointly owned Tube-Wells | | operational Tube-Wells | |
|---|---|---|---|---|---|---|---|---|
| Mixed Irrigation Village (Tohl Kalan) | No | %age | No | %age | No | %age | No | %age |
| Marginal Farmer | 7 | 9 | 5 | 28 | 12 | 29 | 16 | 16 |
| Small Farmer | 22 | 28 | 6 | 33 | 11 | 26 | 26 | 27 |
| Medium Farmer | 32 | 41 | 3 | 17 | 14 | 33 | 38 | 39 |
| Large Farmer | 17 | 22 | 4 | 22 | 5 | 12 | 18 | 18 |
| Total no of wells | 78 | 100 | 18 | 100 | 42 | 100 | 98 | 100 |
| Tube-well Irrigation Village (Gharinda) | | | | | | | | |
| Marginal Farmer | 4 | 3 | 0 | 0 | 0 | 0 | 4 | 3 |
| Small Farmer | 6 | 5 | 0 | 0 | 0 | 0 | 6 | 5 |
| Medium Farmer | 31 | 26 | 0 | 0 | 3 | 100 | 34 | 28 |
| Large Farmer | 77 | 65 | 1 | 100 | 0 | 0 | 78 | 64 |
| Total no of wells | 118 | 100 | 1 | 100 | 3 | 100 | 122 | 100 |
| Tube-well Irrigation Village with Problems of Depletion (Ballab-e-Darya) | | | | | | | | |
| Marginal Farmer | 7 | 9 | 0 | 0 | 4 | 27 | 13 | 14 |
| Small Farmer | 5 | 7 | 0 | 0 | 10 | 67 | 15 | 16 |
| Medium Farmer | 28 | 37 | 0 | 0 | 1 | 7 | 29 | 31 |
| Large Farmer | 36 | 47 | 0 | 0 | 0 | 0 | 36 | 39 |
| Total no of wells | 76 | 100 | 0 | 0 | 15 | 100 | 93 | 100 |

Source: Questionnaire surveys in various villages from May to July, 2009

Table 5. Types of Tube Well Ownerships across Farm Size Classes

Data reveals that joint ownership of wells mostly rests with small and marginal farmers (Table-5). Large farmers mostly have wells under individual ownership. In some cases they consolidate their shares in the wells by purchasing from other shareholders. A positive correspondence is also noted for incidence of joint ownership and groundwater depletion (Table - 5). With depletion, the running cost of groundwater irrigation increases as continuous deepening becomes mandatory to sustain tube-well irrigation. In such situations the joint ownership helps the small and marginal farmers to share the cost and have access to groundwater irrigation. The cost of the well is borne by all the share holders in proportion to the number of shares they own and the proportion of the land they will be irrigating with the help of the shared water extraction machine. In cases where the shareholders don't cover their proportion of the costs, they are excluded from use of the pump set. If a shareholder voluntarily withdraws his share from a joint well the remaining shareholders contribute money to take out his share. The maintenance and deepening of the well is also jointly done by all the share holders.

In reality, however, the cost benefit sharing of the jointly owned wells are much more complex. While the details of the management of jointly owned wells for every case is not documented in detail, but interviews suggest that the incidence of conflict in the process of sharing of water from jointly owned wells is widespread and that practical difficulties surrounding pumping and management of shares and ownerships are of the most important source of conflict which often results in differential access between dominant owners and others who are less capable of exercising their partial ownership rights. Where scarcity is an issue, rights are likely to come in conflict. Conflicts among the shareholders are common regarding the number, spacing and time of the 'turns'[4] in irrigating their respective farms. The disputes are countless during the kharif season when virtual scarcity of water increases with cultivation of rice. Many disputes also arise due to the erratic power supply[5], which disrupts schedules for sharing available pumping time. Village panchayats (informal village courts) are often involved in resolving such disputes but conflicts continue to resurface in the next period of scarcity. Many disputes are only resolved when one shareholder buys the others out. In some cases this is accomplished by poor farmers selling their land along with their shares in a well. In addition disputes often occur over the need to deepen wells. Shareholders with different land holdings disagree regarding the distribution of the benefits from well deepening and one or more refuses to contribute to the cost. There are also instances of cases where wells are abandoned due to prevalence of too many shareholders and the emergence of numerous disputes. Conflicts were even noticed in cases where farmers voluntarily wanted to take out his share for reasons like migrating to urban areas or abroad, changing occupation, buying land somewhere else or even setting up individual well. The shareholders do not agree to pay for the withdrawn share in the joint wells. In such cases, the individual (who wants to leave the partnership) either goes without getting his share paid or sell off his land. Conflicts in crop selection were also common where some shareholders wanted to grow some other crop but could not do so because of the collective decision of the shareholders. In well sharing per person availability of water also declines (especially with incessant falling of water tables), the shareholders have to wait for their turns to irrigate their crop. This reduces the quality of irrigation as both availability and the control over the water supply decline.

While sharing of water from a joint well is often problematic, positive features also exist. The fact that about 62 % of the jointly owned wells are accessed by farmers owning less that 4 acres of land indicates greater groundwater accessibility to the small and marginal farmers through this system. In the villages there are informal rules governing the sharing of costs and benefits from a jointly owned well and village panchayats play a role in redressing disputes. Thus, joint ownership system promotes accessibility to groundwater irrigation and particularly benefits those who can not afford a well of their own because of lack of resource

---

[4] A specific number of hours and a specific time are fixed for each shareholder to use the pump or the tube-well to irrigate his land.
[5] During the peak time of irrigation of rice (May – June) the electricity supply in the villages on an average varies from 6 to 8 hours.

Equity in Access to Irrigation Water: A Comparative Analysis of Tube-Well Irrigation System and
Conjunctive Irrigation System

59

and also due to ownership of small fragments of land. While many joint wells fail due to two interrelated reasons; declining groundwater levels and the lack of finances for well deepening etc., many joint well ownership also become successful in providing groundwater access to small and marginal farmers who join hands in the time of scarcity to jointly harness and share the benefits of this (groundwater) resource which would not have been possible with individual efforts (investments). Many farmers believe that joint ownership of wells for this very reason is a better solution for groundwater accessibility especially in times of depletion but feel that joint ownership among kins and friends do not materialize as their individual small land holdings are spaced at greater distances and since joint ownership requires adjustability and compatibility to avoid conflicts the farmers are not comfortable to become partners of just any (neighbouring plot's) farmer. When the farmers of distant fields become partners in joint wells, disputes commonly arise as many farmers object to passing of irrigation pipes through their plots and mischievous incidences of damaging pipes and disrupting (stealing) water supplies takes place. In such cases when joint ownership of wells fails, they resort to buying water which not only becomes costly but also exploitative at times. While the share system (partially) promotes equity in access to groundwater, depletion reinforces inequality in the village societies where many joint owners become heavily indebted and are eventually forced to sell their shares along with their parcels of land.

## 3. Equity to groundwater irrigation accessibility

To examine the access to the groundwater resource, two parameters, namely, physical and economic access to the resource is discussed. The physical access to resource is the groundwater used by the farmers measured in volume (acre-hours); economic access is the cost per unit volume of water used/accessed. The equity to resource was examined by classifying the farmers in two ways – on the basis of holding size and on the basis of the different agro-ecosystems at different levels of resource depletion. It is evident that physical access to groundwater resource is skewed towards the higher landholding classes (Table- 6). The inequality to physical access to groundwater resource is due to the inequality to land holding sizes. If we negate the land holding factor and work out the physical access realised to groundwater resource on the basis of per unit of holding size for each class, we observe that the groundwater realised per acre of holding size is lowest in the groundwater depleted village of Ballab-e-Darya which indicates towards low yield of tube-wells due to progressive water table depletion. There is also inequality in water accessibility among marginal and large land holdings as farmers of  marginal and smaller land holdings are incapable for chasing water tables as fast as the resource rich farmers. The per acre accessibility of groundwater is almost same among the tube-well irrigation village of Gharinda where since the water table is comparatively at shallower depths, the farmers across all categories can access groundwater.  In the mixed irrigation village of Tohl Kalan the per acre accessibility to groundwater is low for the marginal farmers because most of them (marginal farmers) irrigate with canal water as investment in tube-well for small plots of lands are not economical and with availability of canal water it is also not a mandatory option. The other parameter of equity, the economic access to groundwater, is also more skewed towards the larger land holding groups (Table - 6). Thus on one hand there is worsening physical shortage of water for small and marginal farmers and on the other there is also a scarcity of economically accessible water.

| Particulars | Marginal Farmer | Small Farmer | Medium Farmer | Large Farmer |
|---|---|---|---|---|
| **Mixed Irrigation Village (Tohl Kalan)** | | | | |
| **Total water used across all farms (acre-hour)** | 13208 (3) | 46074 (11) | 141384 (34) | 219460 (52) |
| Water accessed per unit of holding size (acre-hour/acre) | 403 | 593 | 601 | 817 |
| Economic accessibility of groundwater = acre-hour of ground water per rupee of a motorised cost of well* | 48624 | 121872 | 258149 | 831172 |
| **Economic accessibility of ground water per Rs. 1000** | 49 | 122 | 258 | 831 |
| **Tube-well Irrigation Village (Gharinda)** | | | | |
| **Total water used across all farms (acre-hour)** | 4126 (1) | 13128 (2) | 147850 (18) | 646938 (80) |
| Water accessed per unit of holding size (acre-hour/acre) | 515.75 | 625.14 | 634.55 | 652.16 |
| Economic accessibility of groundwater = acre-hour of ground water per rupee of a motorised cost of well* | 64702 | 144432 | 306179 | 745554 |
| **Economic accessibility of ground water per Rs. 1000** | 64.70 | 144.43 | 306.18 | 745.55 |
| **Tube-well Irrigation Village with Problems of Depletion (Ballab-e-Darya)** | | | | |
| **Total water used across all farms (acre-hour)** | 10702.5 (3) | 27558 (8) | 109765 (31) | 210711 (59) |
| Water accessed per unit of holding size (acre-hour/acre) | 365.90 | 314.05 | 442.16 | 565.67 |
| Economic accessibility of groundwater = acre-hour of ground water per rupee of a motorised cost of well* | 21858.81 | 121194 | 227474 | 708611 |
| **Economic accessibility of ground water per Rs. 1000** | 21.86 | 121.19 | 227.47 | 708.61 |

Note: *Figures in parentheses are percentage to total and the cost is calculated as actual running cost incurred if diesel pumps were used*
Source: Questionnaire surveys in various villages from May to July, 2009

Table 6. Equity to Groundwater Irrigation Accessibility for Farm Size Classes

## 4. Equity in net returns from agriculture

To examine the extent of inequity in access to groundwater irrigation, the extent of inequity of net returns per acre realized for different landholding size classes is taken as a proxy variable. Various measures of income inequality were estimated (Table-7) and is also presented in the Lorenz curve (figure-2). Inequality of agricultural return distribution is indicated by the degree to which the Lorenz curve departs from the diagonal line: the further the curve is from the diagonal line, the more unequal is the farm income distribution, and vice versa. For all these measures as well as the Lorenz curve, it can be

Equity in Access to Irrigation Water: A Comparative Analysis of Tube-Well Irrigation System and
Conjunctive Irrigation System

61

| Inequity measures | Mixed Irrigation Village (Tohl Kalan) | Tube-well Irrigation Village (Gharinda) | Tube-well Irrigation Village with Problems of Depletion (Ballab-e-Darya) | Total of all samples |
|---|---|---|---|---|
| Gini concentration ratio (GCR) | 0.070 | 0.008 | 0.218 | 0.099 |
| Theil Entropy index | 0.039 | 0.003 | 0.040 | 0.028 |
| Standard deviation of logarithmic income | 1.006 | 0.204 | 1.687 | 0.966 |
| coefficient of variation | 0.444 | 0.270 | 0.544 | 0.420 |

Source: Authors own calculation

Table 7. Measures of Income Inequality in Different Sample Villages

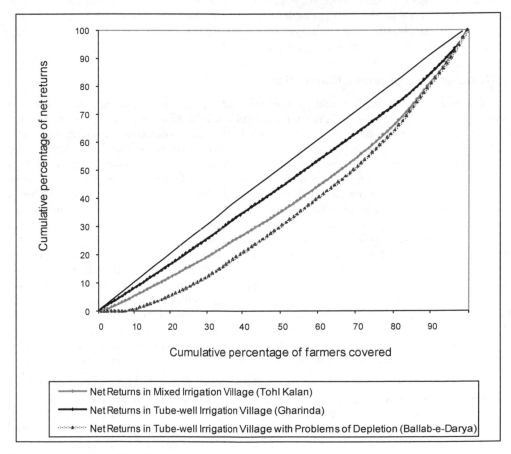

Fig. 2. Distribution of Net Returns to Cultivation

concluded that the net returns realized by farmers using groundwater irrigation in Gharinda is more evenly distributed than and in Ballab-e-Darya where there is problems of groundwater depletion. This is due to the more skewed access (distribution) to groundwater irrigation among the various classes of farmers in Ballab-e-Darya. Only a few marginal and small farmers have access to groundwater in Ballab-e-Darya on account of rising cost due to depletion. However it was not in case of Gharinda where groundwater accessibility was more equal. In Tohl Kalan the less inequality in net returns from agriculture was due to fact that a majority of small and marginal farmers who do not have tube-wells use canal water and have a large number of joint wells to supplement canal irrigation. A high proportion of marginal and small farmers being shareholders in joint wells in Tohl Kalan reduce inequality in resource among the different classes of farmers and thereby to irrigation surplus. But in Ballab-e-Darya due to deeper water tables and progressive receding of water table, the investment costs and maintenance of water yield in wells are very high. So the marginal and small farmers are fearful to go in for new bores on an individual as well as joint basis, thereby limiting their access to the resource. The non existence of any subsidiary source of irrigation other than tube-well irrigation further worsens the inequality in groundwater access and income distribution in Ballab-e-Darya. This shows that groundwater depletion plays a major role in inequitable distribution of groundwater irrigation access in a water scarce region like Punjab.

## 5. Conclusion and policy implication

The study reinforces the fact that growing inequity in access to groundwater leads to a process of continued social differentiation, which results in deprivation, poverty and the consolidation of inequitable power relations within local communities. Declining water levels and overexploitation of groundwater further leads to equity and sustainability problems and deteriorating socio-economic conditions. The immediate consequence of groundwater depletion is linked with the increasing cost of groundwater irrigation in terms of both capital and operating costs which is an increasing function of depth of water table. If the receding water table becomes a common phenomenon, the cost of groundwater irrigation rises in perpetuity. In case of considerable decline in water table, the external effect could not be only extra capital and operational costs but also lower farm output because of either reduced availability of water or lesser use of water at the enhanced cost of lifting it, or both. When the enhanced cost of water lifting exceeds the benefits from the use of such water for small farmers with traditional modes of groundwater irrigation that they are forced to give up irrigated farming altogether. Thus with continuous decline in the water table, the small and marginal farmers get deprived of groundwater or pay higher irrigation charges or they adjust their agriculture operations according to the accessibility of the water which largely depend on the tube-well owners who are generally large framers. This increase cost and severely affects the small farmers' production in the long run.

In the last twenty years gradual increase in groundwater access has undermined maintenance of canal irrigation systems Punjab which is evident from the government statistics which shows net area irrigated by canals has been declining and at present it is less than 27%. Field investigations reveal that the actual area under canal irrigation is further less as most of the canals have dried up and there is hardly any supply of canal water. Lack of maintenance of canal network and declining public investment in canal infrastructure

Equity in Access to Irrigation Water: A Comparative Analysis of Tube-Well Irrigation System and
Conjunctive Irrigation System

63

have consequently led to shrinking area under canal irrigation further compelling the farmers to increasingly depend on groundwater for irrigation. It seems that the subsidy in irrigation has shifted from canal subsidy to electricity subsidy in agriculture in Punjab to the extent that agricultural electricity is free in Punjab. In the process it has shifted the determinants of water access away from communities and into the hands of few resource rich individuals who can invest capital in upgrading water technology and continuously deepen wells with depletion.

This has broader repercussions in the agricultural communities in Punjab. Firstly with an inherent inequality attached to groundwater ownership and accessibility on account of being privately initiated and monitored, the electricity subsidy consequently is disproportionately shared. But with declining water tables (for which large farmers are more responsible as they pump out more water and have large plots of land), the small and marginal farmers lose out on improvising their groundwater technology and competitive deepening and in the process get increasingly excluded form the financial grants (in this case free electricity) given by the government to facilitate the farmers to augment agricultural production. Secondly, when canal water is available in the villages, the small and marginal farmers (can at least) avail of irrigation water from canals or use canal water supplemented by tube-well water even when they do not own groundwater technology which (as of now)[6] is entirely a private initiative to start with and maintain. So in such cases where canals exists, these marginalised farmers can at least use some form of government grant (the canal water subsidy) to augment production (if not the groundwater subsidy) rather than being completely deprived. But the irony is that, the canal water subsidy although exists, due to lack of maintenance, most of the canals have dried out leaving the farmers no option but to depend on groundwater for irrigation. Thirdly, since this (electricity subsidy) financial assistance is not 'targeted' it is (mis)appropriated by the wealthy and does not reach the needy farmers who actually require this support. Lastly, the electricity subsidy is enhancing groundwater depletion which in turn is enlarging the gap between the rich and the poor making the agriculture ecologically unsustainable and socially impoverished in Punjab.

In the absence of surface water irrigation, groundwater withdrawals will tend to outstrip the groundwater recharge, with consequent downward pressure on the water table. In the presence of canal irrigation the pressure on water table eases in two ways: part of the demand for irrigation water shifts to canal water and seepage from unlined part of the canal network augments groundwater recharge. Thus a policy of simultaneous development of surface and groundwater irrigation will prevent permanent decline of water table in arid or semi-arid or low rainfall areas because of over-exploitation of groundwater which in the long run will also lead to sustainale agriculture. Sustainable water management should consider the environmental and equity issues and should cater to the needs of the poor and underprivileged who are generally marginal and small farmers.

## 6. References

Abu-Zeid, Mahmoud (2001), "Water pricing in Irrigated Agriculture", *Water Resources Development*, Vol.17, No.4.

---

[6] As no government tube-wells are functional and no credit is given to install new tube-wells.

Bhatia, Bela (1992), "Lush Fields and Perched Throats: Political Economy of Groundwater in Gujarat", *Economic and Political Weekly*, Vol.27, No.51-52, December 19.

Dhawan, B.D. (1982), "Development of Tube well Irrigation in India", Agricole Publishing Academy, New Delhi.

Dhawan, B.D. (1990), "Studies in Minor Irrigation: With Special Reference to Groundwater", Commonwealth Publications, New Delhi.

Dubash, N. K. (2002), "Tube well Capitalism: and Groundwater Development an Agrarian Change in Gujarat", Oxford University Press, New Delhi.

Gill, Kanwaljit & S.S., Gill (1990), "Agricultural Development and Industrialization in Punjab", *Economic and Political Weekly*, Vol.39, No.13, Nov 10.

Janakarajan, S. (1993), "Economic and Social Implications of Groundwater Irrigation: Some Evidence from South India", *Indian Journal of Agricultural Economics*, Vol.48, No.1.

Janakarajan, S. and Marcus, Moench (2006), "Are Wells a Potential Threat to Farmers' Well-being? Case of Deteriorating Groundwater Irrigation in Tamil Nadu", *Economic and Political Weekly*, Vol.41, No.37, 16 – 22 September.

Jasveen, Jairath (1985), "Private Tubewell Utilization in Punjab: A Study of Cost and efficiency", *Economic and Political Weekly*, Vol.20, No.40, October 5.

Moench, Marcus H. (2000), "India's Groundwater Challenge", *Seminar*, No.486, February.

Moench, Marcus H., (1992), "Chasing the Water Table: Equity and Sustainability in Groundwater Management", *Economic and Political Weekly*, Vol.27, No.51-52.

Nagraj, N. and Chandrakanth, M.G. (1997), "Intra and Inter Generational Equity Effects of Irrigation Well failures", *Economic and Political Weekly*, Vol.32, No.13.

Pant, Niranjan (Ed) (1984), "Productivity and Equity in Irrigation Systems", Ashish Publishing House, New Delhi.

Sarkar, Anindita (2009), "Sustainability Status of Rice-Wheat Cropping Pattern in Punjab: A Comparative Analysis of Different Irrigation Systems" *Environment, Development & Sustainability*, Volume 11, No. 4, Pp. 751 – 763.

Sarkar, Anindita (2011) "Socio-Economic Implications of Depleting Groundwater Resource in Punjab: A Comparative Analysis of Different Irrigation Systems", *Economic and Political Weekly*, Volume 46, No. 7, Pp. 59 – 66.

Shah, Tushaar (1993), "Groundwater Markets and Irrigation Development: Political Economy and Practical Policy", Oxford University Press, New Delhi.

Shah, T. (1988), "Transforming Groundwater Markets in Powerful Instruments of Small Farmers Development: Lessons from Punjab, Uttar Pradesh and Gujarat", Overseas Development Institute, London.

Shah, T. and K.V. Raju (1987), "Working of Groundwater Markets in Andhra Pradesh and Gujarat: Results of Two Village Studies", *Economic and Political Weekly*, Vol.22, No.6, February7.

Shah, Tushaar (1998), "Water against Poverty: Livelihood-Oriented Water Resource Management", *Water Nepal*, Vol.6, No.1.

Siddhu, H.S. (2002), "Crisis in Agrarian Economy in Punjab Some Urgent Steps", *Economic and Political Weekly*, Vol.37, No.18, July 27.

Singh, Chhatrapati (ed.) (1992), "*Water Law in India*", Indian Law Institute, New Delhi

Singh, Dalbir (2002), "Groundwater Markets in Fragile Environments: Key Issues in Sustainability", *Indian Journal of Agricultural Economics*, Vol.57, No.2.

# 4

# Effects of Irrigation-Water Pricing on the Profitability of Mediterranean Woody Crops

M. A. Fernández-Zamudio[1], F. Alcon[2] and M. D. De-Miguel[2]
*[1]Valencian Institute of Agricultural Researchs*
*[2]Technical University of Cartagena*
*Spain*

## 1. Introduction

Tree-crops play a fundamental role in Mediterranean Spanish regions. Irrigated farmlands of citrus and a wide range of fruit-trees are characteristic of the Valencian Community and the Region of Murcia, for instance. These are intensively grown crops whose production is destined for fresh consumption, and are highly competitive due to the large proportion that is exported. However, there are other crops commonly grown in the inland areas which, although not weighty in economic terms, are of major importance. Examples are the irrigated table grapes, olives, almond trees and vines which are grown mostly under dry or rainfed conditions. The latter are paramount in the settlement and permanence of the rural population, especially in regions where agriculture remains the economic mainstay. Furthermore, management is also highly environmentally sustainable, because these crops consume minimum amounts of water. In this chapter these four species have been chosen to represent the Spanish Mediterranean woody crops.

In recent decades, socioeconomic and political changes have led to a clear transformation of the farming communities in the Mediterranean regions. Both the industrial and the service sectors continue putting strong pressure on labour resources, especially among young people, which limits generational take-over and leads to the disappearance of family farms, giving little incentive to continue farming activities.

The climate in these Mediterranean areas is characterized by short, mild winters, hot summers and, in many areas, with an annual precipitation below 300 mm, making water the scarcest and most valuable natural resource in these regions. Thus, the aquifers are subject to excessive extraction, often resulting in poor water quality. Main crop production can only be guaranteed with some type of irrigation. Official markets of irrigation water have not been developed yet, but in producing regions the resource is exchanged; it is especially common that growers administer their allocation and divert it to more profitable plots. This propitiates marginal management, where a plot is maintained during a period with only basic cultural practices.

To encourage a sustainable use of water in agriculture, the recently entered into force European Water Framework Directive (WFD) proposes the full cost recovery related to water services under the polluter pay principle. For this end, a water pricing policy should be established by Member States and, it is foreseeable that the price of irrigation water will increase in such a way that the final price of irrigation water to be paid by the grower will cover all of the costs (economic and environmental) incurred in delivering it. But the field prices of most agricultural products are increasingly lower, and any increase in production costs may seriously affect its feasibility.

In this context, the objective of this study is to analyze the potential trend of certain Spanish Mediterranean woody crops, on the establishment of tariff policies that differ from those currently implemented. To achieve this goal, the water demand curves obtained for olives, vineyard, almond trees and table grapes have been analyzed. The results will enable policy makers and water resource managers to be informed of the effect that the introduction of a pricing policy would have on Mediterranean agriculture influencing decision making, taking into account the environmental objectives of the WFD while maintaining the social and environmental benefits of traditional farming in the Mediterranean regions.

The remainder of the chapter is organized as follows. In the next section, the economic analysis of irrigated water use is reviewed, as well as the main methodological approach. In sections 3 and 4 two case studies are analyzed. Finally, general conclusions are drawn.

## 2. Analysis of irrigation water pricing on Mediterranean crops

### 2.1 Background

Water is becoming an increasingly scarce resource in the world due to population growth, improved quality of life and diminishing resources. Thus, efficient allocation of water becomes increasingly important and must be achieved if we are to reconcile economy and society, because water management is often guided by contradictory aims. Furthermore, to achieve efficient water management, given the competitiveness of demand, it will be necessary to consider the distribution of wealth among different users, and ensure the right to access to water following environmental criteria that enable ecological sustainability. Therefore, those responsible for water management require broad vision, taking into account the results of different water management policies, and also to learn how to select the specific management tools for each situation.

Numerous papers are found in scientific literature that determine the economic effects of the application of different water policies or the adoption of one type of irrigation system or another. Works related with irrigation water management in Mediterranean crops are: Gurovich (2002) studied the energy cost of irrigating the Chilean table grape; Hearne and Easter (1997) analyzed the impact of applying water markets to the cultivation of fresh grapes; Bazzani et al. (2004) investigated the effects of putting into practice the Water Framework Directive in Europe; Jorge et al. (2003) carried out an economic evaluation of the consequences of drought on the Mediterranean crops; and Fernández-Zamudio et al. (2007) obtained irrigation water demand curves for the Spanish table-grape.

The WFD sets out a strategy for general water management and establishes environmental objectives for water bodies in order to achieve a so-called "good ecological status". To achieve these objectives in innovative ways, the WFD prescribes using economic tools and principles, among which is the principle of full cost recovery related to water services. The incorporation of this principle will lead to member states restructuring existing water rates to allow recovery of full costs of water services, including both the environmental and the resource. In this sense, water pricing policies are viewed by the WFD as a way to internalize the costs so they can educate users about water shortage.

Economic theory sets that farmers would respond to an increase in water prices by reducing their consumption, in accordance with a negative slope demand curve. Thus, the implementation of water pricing measures would create incentives to use water efficiently and, in accordance with the WFD, would contribute to the achievement of environmental objectives. It is believed that the use of suitable water pricing policies would encourage users to limit their water use (Easter and Liu, 2005).

However, the tariff policy applied must consider the specific conditions of each irrigation area, since its effect will be very different depending on their characteristics. Specifically, in south-eastern Spain, where the irrigation water demand curve is usually inelastic, a tariff policy would be valid only from the standpoint of cost recovery, but is not expected to be so from the water savings perspective (Sumpsi et al., 1998; Varela et al., 1998).

Up to our knowledge, the impact of the water pricing policy in Spain was firstly analyzed in three irrigated areas by Berbel & Gómez-Limón (2000). They applied a simple lineal programming model to analyze the impact of using a volumetric water policy instead of an area pricing scheme on agricultural production. They found that demand curve become inelastic and inefficient for a water prices higher than 0.15 €/m³, concluding that volumetric water pricing would have undesirable consequences over the farm income and the employment, reducing also the range of crops available.

Two years later, after WFD entered into force, Gómez-Limón et al. (2002) study both the impact of the cost recovery principle proposed by the WFD and the effect of the common agricultural policy on agricultural irrigation systems using multi-criteria techniques (multi-attribute utility theory). They used an irrigation community, placed in the north-central of Spain (*Bajo Carrión*, Palencia), and they found that the inelasticity of the water demand curve is reached from 0.09 €/m³. From this value it is concluded that an increase of water price would not influence the amount of water demanded. Also, in this irrigation community, Gómez-Limón & Berbel (2000) derive water demand function using weighted goal programming approach obtaining that water pricing policies are not a satisfactory tool for significantly reducing water consumption in agriculture. They argued that water consumption is not reduced significantly until prices reach such a level that farm income and agricultural employment are negatively affected.

In this line, and using a multiattribute utility theory (MAUT) mathematical programming models, Gómez-Limón & Riesgo (2004a) analyzed the impact of the hypothetical implementation of recovery of costs water pricing proposed by the WFD on three homogeneous groups of farmers in the irrigation community *Virgen del Aviso*, in the north of Spain. They found again that the effect of irrigation water pricing vary significantly depending on the group of farmers being considered, being the demand curve of the group with higher ability to pay for the water inelastic from 0.12 €/m³. The same methodology was also applied in an irrigation community of the *Pisuerga channel*, in the north of Spain by Gómez-Limón & Riesgo (2004b) obtaining similar findings.

In general, the works quoted above analyze irrigation water demand curves in irrigation communities of Spain, and all of them have been applied to irrigated areas with a predominance of extensive arable crops. These crops, which represent a majority in Spain, have a yearly production cycle and their production margins are very similar. These characteristics contribute to the replacement of one species by another depending on the harvests. However, it is expected that woody crops show different behaviour than extensive crops due to the higher investment costs and the time needed to recover such investment. Recent studies, such as that by Mesa-Jurado et al. (2010), determine the marginal value of irrigation water in a woody crop, like the olive tree, in a sub-basin the Guadalquivir River Basin using production function based on field experiments. Net marginal values of water obtained from the marginal benefit curve (having deducted the variable costs of production including harvesting and irrigation) were 0.60 and 0.53 €/m³ for an allocation of 1,000 and 1,500 m³/ha respectively.

For the whole Guadalquivir River Basin, a wider analysis of the demand of water has been estimated by all irrigated crops, extensive and perennial. This work has been carried out by Berbel et al., (2011) using the Residual Value Method. This technique is based on the idea that a profit-maximizing firm will use water up to the point where the net revenue gained from one additional unit of water is just equal to the marginal cost of obtaining the water. An approximation to the demand curve can be obtained plotting the average residual water values aggregated for all crops by area and the water consumed by this area. In this work, where citrus and olive tree are considered jointly with extensive crops, the residual water value is estimated between 0.01 and 0.68 €/m³. The high residual water values are explained for the existence of woody crops.

Thus, from the analysis of the previous works it is possible to identify that the demand of water for wood crops seems more inelastic than for extensive crops, due to for small allocation the price is highly valuable, being the marginal water prices of woody crops considerable higher. Studies that analyze the establishment of tariff policies on woody crops are still scarce, and more research based on this kind of crops would contribute to fill this gap of knowledge.

## 2.2 Methodology for analysis of irrigation water pricing

In order to deduce the effects that irrigation water availability and price have on the viability of the main rainfed tree crops grown in Spain as well as table grapes and the woody crops representative of irrigated farmlands in the Mediterranean region, several mathematical calculations have been used in this work. Together with the calculations of shadow prices and irrigation water demand curves, the maximum price that farmers can afford for water resources has been obtained.

The shadow price of irrigation water is the value corresponding to the opportunity cost of having one more cubic meter of water on the farm. The economic impact that is expected from having more water available, for which the approximate value of irrigation water is cited as Euros per cubic meter. This price cannot be considered the actual price that the farmer may pay for the water resource and it must be inferred from the demand curves.

To calculate shadow prices a quantitative analysis has been carried out using the Compromise Programming (CP) technique belonging to the multicriteria paradigm.

Usually, the analyzed objectives in multicriteria models are in conflict with each other, and therefore, to optimize the different objectives simultaneously and achieve an ideal solution is impossible. However, it is possible to determine the small group of effective points that bring us closest to that ideal, which would be the solution in which all the objectives reach their optimum value (Romero & Reheman, 2003). The mathematical essence of this calculation was established by Zeleny (1973) and Yu (1973), and a number of authors have used this technique in agriculture. For example, Sabuni & Bakshoudeh (2004) used the CP to determine the opportunity cost of water on farms, or Ballestero et al. (2002) who analysed the establishing of water markets.

The CP is one of the most commonly applied multicriteria techniques due to its high operativity (Romero & Reheman, 2003), and it is associated with the concept of distance, though not in the geometric sense, but rather the distance or degree of proximity from the ideal. This distance $(d_j)$ of the objective $f_j(x)$ with respect to the ideal $f^*_j$, will be written:

$$d_j = \left| f_j^* - f_j(x) \right| \tag{1}$$

Normally the objectives have very different absolute values or they are measured in different units, therefore before adding the degree of proximity they could all have, one must carry out a dimensional homogenization, giving:

$$d_j = \frac{\left| f_j^* - f_j(x) \right|}{\left| f_j^* - f_{*j} \right|} \tag{2}$$

where $f_{*j}$ is the worst value of the objective when it has been optimized separately and called the anti-ideal value.

Likewise, in the calculation, one must also consider the preferences that the decision centre can show for each objective, for this reason a weight $w_j$ is included. All this means that the effective solutions that come closest to the ideal are achieved by resolving the following optimization problem:

$$Min\ L_p = \left[ \sum_{j=1}^{n} w_j^p \cdot \left| \frac{f_j^* - f_j(x)}{f_j^* - f_{*j}} \right|^p \right]^{1/p} \tag{3}$$

Subject to $x \in F$, where $x$ are the decision variables, $F$ is the set of restrictions of the model, $n$ is the number of the objectives introduced in the modelization and $p$ the metric (Romero & Rehman, 2003).

The points that fall closest to the ideal (called the compromise set) can be bounded between the metrics one and infinite, in other words $L_1$ and $L\infty$ (Yu, 1973), which is considered acceptable, even though there are more than two objectives. The economic significance of these solutions is connected with the traditional optimization based on utility functions and $L_1$ indicates the value of greatest efficiency, while $L\infty$ is the solution with greatest equity (Ballestero & Romero 1991).

For the demand curves, the Multiattribute Utility Theory (MAUT) has been widely founded. The work by Keeney & Raiffa (1976) is a starting point of the MAUT. In essence it consists of being able to establish a mathematical function $U$, which encompasses the utility resulting from a series of attributes, which are previously considered according to the importance each of them has for the decisor. This theory starts from strict mathematical requirements; however, the works by Edwards (1977) or Huirne & Hardaker (1998) show that, although these are not strictly satisfied, one can obtain utility functions that are extremely close to the true utility.

In order to estimate the additive utility functions, the framework developed by Sumpsi et al. (1996) and Amador et al. (1998) has been followed, and later applied by Gómez-Limón et al. (2004). First one calculates the pay-off matrix, and then resolves the following system of $n+1$ equations:

$$\sum_{i=1}^{n} w_i f_{ji} = f_j \qquad (4)$$

for $j = 1, 2, ..., n$ and $\sum_{i=1}^{n} w_i = 1$

With $n$ being the number of objectives considered, $w_i$ are the weights of the different objectives (and therefore, unknown), $f_{ji}$ are the elements of the payoff matrix, corresponding to the values reached by the objective of column-$i$ when the objective of row-$j$ is optimized. Finally $f_j$ is the value of the $j$-th objective in accordance with the distribution of the crops observed.

If the above system of equations has a non-negative solution, then $w_i$ indicate the weights of the different objectives, but this is not usually the case, as there is no set of weights that reproduce the farmers preferences with precision. To approximate the said solution as far as possible, one minimizes the sum of $m_j$ and $p_j$ , for which the following lineal program is resolved:

$$Min \sum_{j=1}^{n} \frac{m_j + p_j}{f_j} \qquad (5)$$

subject to:

$$\sum_{i=1}^{n} w_i f_{ji} + m_j - p_j = f_j \quad \text{for } j = 1, 2, ..., n$$

$$\sum_{i=1}^{n} w_i = 1$$

with $m_j$ the variable of negative deviation and $p_j$ the variable of positive deviation. According to Dyer (1977), the weights obtained previously, coincide with the following expression of the utility function, which is separable, additive and lineal for each attribute $f_i(x)$,

$$U = \sum_{i=1}^{n} \frac{w_i}{k_i} f_i(x) \qquad (6)$$

where $k_i$ is a normalizing factor, for instance the difference between the best or ideal value for each objective, $f_i^*$, and the worst or anti-ideal, $f_{i*}$, which are extracted from the pay-off matrix, with the additive utility function finally being expressed as:

$$U = \sum_{i=1}^{n} w_i \frac{f_i(x) - f_{i*}}{f_i^* - f_{i*}} \qquad (7)$$

The utility function it is assumed that the farm owners will maintain their psychological attitude with regard to decision taking for a short to medium term. From this point the study goes on to look at a series of simulations with rising prices of irrigation water, in such a way that, each price is a new scenario in which utility is maximized, and from which a cropping plan is derived with a specific demand for irrigation water.

## 3. Case 1: Rainfed Mediterranean tree crops (olive, vineyard and almond): Response to variations in irrigation water pricing

Spanish Mediterranean dry-farming is predominant in the large inland extension, and the most traditional and characteristic crops are the olive (*Olea europea*), the vineyard (*Vitis vinifera*) and the almond (*Prunus dulcis*). All of them are shared in the farms in different proportions. These crops have helped in maintaining the countryside, which is one of the marks of the cultural identity of these regions, protect the soil from erosion, and they can also be considered an important promoter of human activity.

In the arid regions of the Mediterranean, agriculture is strongly conditioned by the irregularity of the climate, specially the rains. The most important natural resource is water, which is in short supply and of the greatest value. Consequently, the availability of water for irrigation significantly increases cropping yield, assures a greater regularity in harvest, and decreases the economic risks in farming.

The size of the farm also exerts an influence, but in contrast to the small-holding structure that is characteristic on the coast, in these inland regions the land is not considered as such a restrictive factor, and normally, it is rather the lack of family labour needed to cultivate the farm in optimum conditions that leads to marginal management. "Marginal management" defines non-definitive semi-abandonment, in which these three tree crops survive left to the mercy of the climate. This is often the case in the regions under study and at specific periods of time when, due to the lack of labour or profitability, the farmers do not optimize crop care and limit it to a minimum (Fernández-Zamudio et al., 2006).

### 3.1 The most outstanding traits of the dry regions

The present study focuses on the Valencian Community, where dry-lands account for 56% of this area. Here the olive, vineyard and almond are the most extensive crops, representing 35% of the worked lands in this region (CAPA, 2011). These three crops can be cultivated in strict dry-farming or with irrigation at very specific moments (irrigated relief), considered as essential to ensure harvest, and in the case of the vineyard and olive conventional drip irrigation is also possible, with greater and more continuous flow.

In order to determine the optimal cropping plans in the Mediterranean cropping dry-land, the following objectives have been chosen: one of economic nature (to maximize profits), another social (minimization of total annual workforce) and the other environmental (minimization of irrigation water consumption). The mathematical expression of these three objectives is:

$$Max \sum_{i=1}^{n} NM_i \cdot X_i \tag{8}$$

$$Min \sum_{i=1}^{n} Q_i \cdot X_i \tag{9}$$

$$Min \sum_{i=1}^{n} TL_i \cdot X_i \tag{10}$$

Where $NM_i$ is the net margin of the activity $i$, $X_i$ is the surface area, $Q_i$ is the annual irrigation water supplied and $TL_i$ is total labour employed annually.

### 3.2 Information and methodology

Family farms predominate in the Mediterranean inland regions; therefore, the analysis is carried out choosing a representative farm, with 32 hectares of land and a full-time family Agricultural Work Unit (AWU). Within the Valencian Community, the study was located in the *l'Alcoià* area, in Alicante province. This is a region with a semi-arid climate, with the risk of frost from November to March, and 474 mm of average rainfall per year. In this region there are two zones that are very different in terms of slope, but the risk of natural erosion is considered moderate. The irrigation water mainly comes from private wells, and a small proportion distributed by the Irrigation Community of the River *Vinalopó*.

The calculations have been applied to two modelization scenarios that are real in these regions, and the differences are exclusively in the degree of mechanization existing on the farm. In the "manual-scenario" low-powered mobile equipment was used together with traditional harvesting and hand-picking, and in the "mechanized-scenario" higher-powered mobile equipment is considered.

In this study the decision variables, or unknowns of optimization, are the surface area in farm for each crop-growing activity. Olive, vineyard and almond, which are most characteristic of the region, have been introduced, being the main difference the amount of irrigation water (Table 1). To calculate the net margin of an activity, the variable costs, which include hired labour and the fixed costs are subtracted from the income earned through selling the production -together with the subsidy if there is one. In calculating the income, the average production for each crop-growing activity was fixed according to the data taken in the region and after validating them with experts. In the production figures, considered inter-annual variability in yield is recorded. For the prices, the average values have been taken as those perceived by the farmers, according to official statistics (CAPA, 2011).

To bring the models closer to the real conditions in the region, a number of restrictions have been taken into account, and have been introduced equally in both scenarios:

-   Crop area: A total of 32 hectares are available on the farm.
-   At maximum 30% of the available surface area can be subject to marginal management, concept already defined above.

- Given these are woody crops, and that the models under consideration are static and short-term, the maximum surface area of each species is limited to its present value (32% in olive, 8% in almond and 60% in vineyard). This restriction permits changes in variety within a species, changes in the type of irrigation, or for this to pass to marginal management.
- With respect to the availability of irrigation, due to the dryness of these regions, very strict conditions are introduced, fixing a use equivalent to the levels of habitual consumption, which according to the criteria of the experts consulted, can be maintained medium term (Table 1). Therefore, it is established that only 10% of the available surface area can receive some kind of irrigation, the water supplied cannot exceed 600 $m^3$ monthly for the whole farm, and that the total amount allotted to the farm is of 5,000 $m^3$ annually. The current price of irrigation water is 0.15 €/$m^3$.
- The other restrictions are derived from manual labour. The availability of family labour is fixed at an agricultural work unit (an AWU to be 2,160 hours a year), and hired labour is limited to complement what cannot be covered by family on a three-monthly basis.

The curves obtained will be consequence of the adaptation of the farm to increasing prices for irrigation water in the short term. The simulation models are applied to the manual scenario and to the mechanized scenario and are similar to those used to obtain the shadow prices, with the following considerations:

- The MAUT is obtained for the objectives: maximization of the net margin of the farm and minimization of total workforce.
- From the previously calculated margin (Table 1), the cost of the water (corresponding to the usual price, 0.15 €/$m^3$) is deducted and the value corresponding to each simulation, starting from 0 €/$m^3$, is added.
- The restrictions to the models, and the average volumes of irrigation applied to each variety of crop coincides with those previously described (Table 1).

The utility function characteristic of a representative farm of the region under study has been found, assuming that the farm owners will maintain their psychological attitude with regard to decision-taking for a short to medium term. A series of simulations are made with rising prices of irrigation water, in such a way that, each price is a new scenario in which utility is maximized, and from which we derive a cropping plan with a specific demand for irrigation water. Finally, the set of simulations carried out serves to set out the demand functions and will be a consequence of the adaptation of the farm short-term at increasing prices for irrigation water.

## 3.3 Results and discussion

Typically, the demand curves of the areas placed inland of the Mediterranean coastal, show large inelasticity; therefore farmers display a strong willingness to pay for each cubic meter of water. The question it should ask, then, is whether these high prices are affordable by Mediterranean farmers. For this reason, it has also obtained the maximum price of irrigation water that guarantees the family farm income.

Table 2 shows the achievement levels for each of the three objectives for the metrics $L_1$ and $L\infty$, the different cropping plans and the requirements of hired manual labour. The results

| Species | Varieties, description | Irrigation | Annual water supply (m³/ha) | Net margin manual-scenario (€/ha) | Net margin mechaniz.-scenario (€/ha) |
|---|---|---|---|---|---|
| Olive | Authochthonous: Grossal | Dry-land | 0 | 315 | 314 |
| Olive | Authochthonous: Grossal | Irrigated relief | 700 | 662 | 734 |
| Olive | Authochthonous: Grossal | Irrigated | 1500 | 1314 | 1511 |
| Olive | New: Arbequina | Dry-land | 0 | 441 | 300 |
| Olive | New: Arbequina | Irrigated relief | 700 | 933 | 858 |
| Olive | New: Arbequina | Irrigated | 1500 | 1585 | 1611 |
| Almond | Authochthonous: Comuna Group | Dry-land | 0 | 236 | 330 |
| Almond | Authochthonous: Comuna Group | Irrigated relief | 700 | 151 | 329 |
| Almond | New: var.Late-flowering | Dry-land | 0 | 451 | 577 |
| Almond | New: var.Late-flowering | Irrigated relief | 700 | 360 | 568 |
| Vineyard | Monastell in tube | Dry-land | 0 | 550 | 564 |
| Vineyard | Monastell in tube | Irrigated relief | 1100 | 783 | 807 |
| Vineyard | Monastell in tube | Irrigated | 1900 | 1458 | 1492 |
| Vineyard | Monastell in espalier | Dry-land | 0 | 499 | 655 |
| Vineyard | Monastell in espalier | Irrigated relief | 1100 | 788 | 955 |
| Vineyard | Monastell in espalier | Irrigated | 1900 | 1409 | 1602 |

Source: own calculations

Table 1. Rainfed tree crops, decisional variables: description, net margin for scenarios and annual water supply

show a number of advantages on moving from the manual scenario to the mechanized scenario. With regard to the water requirements, in the most balanced cropping plan ($L_\infty$), which is the one that demands most irrigation, it does not exceed 2,227 m³/year on the whole farm in the manual scenario, and 2,418 m³/year for the mechanized scenario. Given the strict conditions used to establish the models, it is possible to think that the proposed plans will be sustainable, even in these arid agricultural conditions. Moreover, in solution $L_1$ there are plans that do not necessitate irrigation, verifying the continuity of traditional dry-farming on its own.

Having reached this point, it is especially interesting to reflect on the behaviour of the profit maximization objective with respect to minimizing irrigation water. If the compromise sets are calculated just for these two objectives, these points can be represented on a Cartesian plane and, the slope of the line joining the points $L_1$ and $L_\infty$, or trade-off, show us the opportunity cost or the shadow price of the irrigation water, understood in its marginal values. In other words, as the increase in the net margin of the farm if one applies an additional unit of water (Florencio-Cruz et al., 2002).

The lines obtained in both scenarios are represented in Figure 1. The volume of irrigation water required by such a plan is represented on the axis of abscissas, while the axis of ordinates shows the net margin this plan generates. The result is that, the shadow price of the water is 0.76 €/m$^3$ in the manual scenario and 0.87 €/m$^3$ in the mechanized ones. The highest shadow price obtained in both scenarios is very significant, and they are a useful orientation to the value of water in this dry-farming system. However, they have been obtained by exclusively evaluating the impact of irrigation water on the net margin of the farm; therefore, to obtain more rigorous information about how these crops would behave in the event of an increase in water prices, the demand curves will be calculated below.

| | | Manual scenario | | Mechanized scenario | |
|---|---|---|---|---|---|
| | Metrics: | $L_1$ | $L_\infty$ | $L_1$ | $L_\infty$ |
| **Value of objectives:** | | | | | |
| Net margin (euros) | | 10,950 | 14,889 | 12,520 | 16,241 |
| Water (m$^3$) | | 0 | 2227 | 0 | 2418 |
| Total manual labour (hours) | | 2,014 | 2,653 | 1,200 | 1,505 |
| **Optimisation activities (%):** | | | | | |
| Olive. Grossal. Dry-land | | | | 2 | 13.3 |
| Olive. Arbequina. Dry-land | | 32 | 27.4 | | |
| Olive. Arbequina. Irrigated | | | 4.6 | | 5 |
| Almond. Var.Later flowering. Dry-land | | 8 | 8 | 8 | 8 |
| Vineyard. Tube. Dry-land | | 30 | 42.7 | 60 | 60 |
| Farm surface with marginal management | | 30 | 17.3 | 30 | 13.7 |
| **Total hectares properly cultivated (ha)** | | **22.4** | **26.5** | **22.4** | **27.6** |
| **Net margin (euros per cultivated hectare)** | | **489** | **562** | **559** | **588** |
| **Requirement annual of hired manual labour (hours)** | | **631** | **1052** | **0** | **106** |

Table 2. Cropping plan and results for the three objectives analysed for compromise solutions in Spanish dry-lands. (Data for a family farm with an agricultural work unit and 32 hectares)

The demand curves obtained are shown in Figure 2. In the event of applying a hypothetical pricing policy, the way the farm behaves will vary according to its degree of mechanization, although we observe that water consumption in the lowest price range is equal in both scenarios. Such inflexible behaviour of the different price ranges should be highlighted, and undoubtedly the great shortage of this resource in these regions. The high productivity of water, even in small amounts, mean that the price the farm can pay for irrigation water can be increased.

In the manual scenario, there is a first range of maximum demand, between 0 and 0.51 €/m³; it continues with a drop to half the demand for tariffs of 0.52 to 0.55 €/m³ and ends up with cropping plans in completely dry-farming when the water costs over 0.56 €/m³. In the mechanized scenario the demand is constantly at maximum until it reaches 0.91 €/m³, at which point the chosen cropping plan changes to one that is strictly dry-farming. The different response must be looked at in the different degree of mechanization. Technology improves management and enables farms to face more effectively the greater labour requirements that arise from irrigated crops. This limitation is accentuated if the labour (especially harvesting) is carried out manually, and for this reason the mechanized farms are better able to pay higher water prices.

The price of water has repercussions on the cropping plan resulting from each simulation. When the prices are low, water is demanded for irrigation and this is destined solely to the olive, specifically to the olive Arbequina irrigated, but for both the almond and the vineyard dry-farming is always chosen. In the manual scenario the dry farmed olive is the Arbequina, while in the mechanized scenario is the autochthonous variety Grossal. With respect to the vineyard, in the manual scenario it is trained in tube, while in the mechanized it is trained in espalier (which is more productive but requires a greater initial investment). The almond chosen is a late-flowering variety in both scenarios. The cropping plans in the mechanized scenario are more economically viable, which means that, with prices of over 0.51 €/m³, they can demand greater quantities of water than in the manual one.

The main conclusions are:

- The different sections of the demand curves demonstrate very inflexible behaviour, which is justified by the fact that the olive, vineyard and almond are woody species that use small amounts of irrigation water very effectively.
- The effect of the price of water on the farm's income make a higher degree of mechanization necessary in order to face high irrigation-water prices. The current price of water is 0.15 €/m³, and although it could increase, to ensure a minimum income for the farm of 21,500 € annually, water cannot exceed more than 0.24 €/m³ in a manually worked farm, or more than 0.44 €/m³ if it is mechanized.
- To increase mechanization may be the most straightforward strategy to ensure the survival of the farms in the Spanish dry-lands, short to medium term, and likewise strengthen their sustainability. As all the results have been obtained considering the operations that demand the most expensive machinery to be hired, it can be deduced that it is a strategy that can be assumed by all the farms. Thus, it will be essential to increase the degree of mechanization in order to guarantee the viability of this agriculture if the current trend of increasing irrigation-water prices is consolidated.

## 4. Case 2: Irrigated woody crops (table-grapes): Response to variations in irrigation water pricing

### 4.1 The most outstanding traits of the table-grape regions

Spain, with 15.2% of the vine-planted surface area in the world, is the leader in terms of the extension this crop covers. Although only 5% of this production is destined to fresh-fruit consumption, cultivation of the table-grape (*Vitis vinifera* L.) is important given its long tradition. Spain is the fifth table-grape producer world-wide in the northern hemisphere,

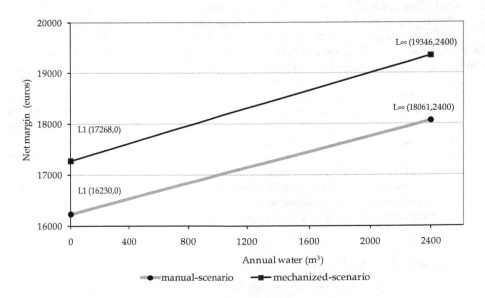

Fig. 1. Shadow price of irrigation water for two scenarios in Spanish dry-lands (Data for family farm with 1 AWH and 32 hectares)

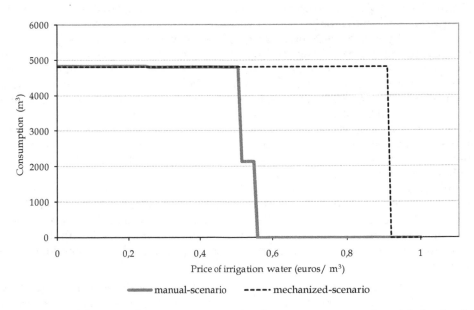

Fig. 2. Demand functions for irrigation water for the two scenarios in Spanish dry-lands

and the second in Europe, surpassed only by Italy. It is the sixth exporter in the world, coming after Chile, Italy, USA, South Africa and Mexico. In 2009, Spain exported 117,143 Tn of fresh grapes, of which 99% were destined to European countries. Table grape is a very typical Mediterranean crop, accounting for 79.3% of the area and 88.8% of the production between the Regions of Murcia and the Valencian Community (MARM, 2011). Both these regions are situated next to the Mediterranean, have a warm climate, but are greatly lacking in rainfall, with average precipitation of below 300 mm annually. Water is the scarcest and most valuable natural resource, given the over-exploitation of aquifers or its low quality, and is the main reason why this crop is partially abandoned. The temporal marginal management are common in table-grape, that is to say, they are left to basic care for a certain amount of time, which habitually consist of maintenance pruning and minimum tillage. If the circumstances that have favoured this situation (scarcity of irrigation water, lack of personal and economic incentive of the agricultural producers, etc.) are longer lasting, then this abandonment becomes definitive. Rejecting optimum crop management is a common practice in these regions of production, and faced with the lack of irrigation water, the choice is to destine the one available to the most economically viable plots.

## 4.2 Information and methodology

The Spanish table-grape farms are very small in size, 72% of those in Alicante (Valencian Community) and 66% of those in Murcia are smaller than 5 ha (INE, 2011). In this study, a 5 ha family farm has been used as reference. The two cropped areas are *Valle del Vinalopó* (Alicante) and *Valle del Guadalentín* (Murcia), geographically closed but technically and managerially different.

Thus, in the *Valle del Vinalopó* there are 9,500 ha of grape (MARM, 2011), forming a well-defined agrarian system that has remained very stable over time. Its most defining feature is its "bagging", by which the bunches of grapes are covered by a paper bag, from just before veraison up until harvesting. Technology is found at an acceptable level, but it is possible to improve the mechanization of the labour. The varietal composition has undergone scarce variation over time, and is fundamentally based on the Italia and Aledo varieties, although important changes are foreseeable in the coming years, and seedless varieties will be introduced, which are still in minor representation. The situation is different in the Region of Murcia, where the surface area dedicated to grape has increased in recent years, reaching 5,159 hectares in 2010 (CARM, 2011) and where new production zones have appeared, with large business producers and a strong bid for seedless varieties. In general terms one can also talk about family farms, but the important investment in capital and technology mean that noticeable differences exist between Murcia and Alicante. There is intense activity concerning the introduction of new plant material, the traditional varieties like Italia and Ohanes have diminished, and instead there has been an increase in the surface area planted with Dominga, Napoleon and, above all, the seedless or early varieties (Superior, Crimson and Red Globe for example).

Currently, in both regions, the growing operation being perfected concerns particularly irrigation, which is bringing about massive implementation of drip irrigation. This type of irrigation covers 50% of the surface area at present, and it is foreseeable that it will reach

90% soon. The shortage of irrigation water in these zones, the irregularity of the supplies and the deficiencies in the quality also encourage the grape-growers to construct accumulation reservoirs, which is more widespread in the Murcia region.

Therefore, the technological improvements that are being adopted more quickly in Spanish table-grape cultivation are: an increase in the average power of the machinery to carry out the labour and the phyto-sanitary treatments, use of tying machines for the summer pruning, use of pre-pruners (in espaliers), generalized use of shredders for the pruning remains and substitution of the traditional irrigation systems for programmed drip irrigation. Moreover, they are improvements that are beginning to spread to the new staking structures (higher espalier in Y in Alicante), and the use of mesh or plastic covering, which are common in Murcia (Fernández-Zamudio et al., 2008).

Maximizing profits (equation 8) is the primary objective of Spanish grape growers, but given the strict water conditions of the regions studied, they must also minimize consumption of irrigation water (equation 9) The decisional variables will be the surface areas occupied by the different growing activities. Table 3 describes the modelized variables, together with their annual water allotment and the net margins in the two technological scenarios analyzed. Scenario-1 represents the traditional production conditions for table-grape in the two zones. From these, one moves to another productive context, in which a series of technological improvements have been adopted, towards a scenario where the two zones would appear to be moving towards according to regional agricultural technicians (scenario-2). The restrictions of the models are a maximum monthly and yearly irrigation allotments, maximum area of cultivation that will be subject to adoption of drip irrigation, change to Y trellises and covering with net screening, as well as those restrictions derived from the market (new varieties introduction).

For analyzing the two objectives together and obtain water shadow prices, compromise programming has been used, while to obtain the demand curves a lineal mathematical programming has been applied, being the objective:

$$Max \sum_{i=1}^{n} NM_i \cdot X_i - Q_i \cdot X_i \cdot p_q \qquad (11)$$

Where $NM_i$ is the net margin of the activity $i$, $X_i$ is the cultivated area, $Q_i$ its yearly quota of irrigation water. Also, $p_q$ is the price of irrigation water for each parameter (from zero to 4 Euros per cubic meter). This is the real price that the grower pays for each cubic meter of water; this price includes administrative costs of delivery, maintenance of the infrastructure, energy for pumping, and other taxes or charges. The restrictions of the model are the same as in compromise programming.

## 4.3 Results and discussion

From the optimal cropping plans obtained (Table 4), it is possible to deduce that technological improvements will broaden economic expectations of grape growers. Net profits will increase by a mean of 15% in Alicante and 97% in Murcia, figures that concord with technological changes introduced in each region.

| (1) Variable description | Irrigation type | Scenario | $Q_i$ (m³/ha) (2) | Net Margin (3) | |
|---|---|---|---|---|---|
| | | | | Scen.-1 (€/ha) | Scen.-2 (€/ha) |
| A Aledo.Traditional espalier. Bagged | Flood | 1& 2 | 3900 | 7311 | 7430 |
| A Aledo. Traditional espalier. Bagged | Drip | 1& 2 | 4000 | 8156 | 8015 |
| A Aledo. Y espalier. Bagged | Drip | 1& 2 | 4000 | 8999 | 8836 |
| A Italia. Traditional espalier. Bagged | Flood | 1& 2 | 3900 | 5000 | 4917 |
| A Italia. Traditional espalier. No bagged | Flood | 1& 2 | 3900 | 5233 | 5150 |
| A Italia. Traditional espalier. Bagged | Drip | 1& 2 | 4000 | 4916 | 4872 |
| A Italia. Traditional espalier. No Bagged | Drip | 1& 2 | 4000 | 5638 | 5582 |
| A Italia. Trellis. Bagged | Flood | 1& 2 | 3900 | 5199 | 5247 |
| A Italia. Trellis. Bagged | Drip | 1& 2 | 4000 | 6160 | 6223 |
| A Italia. Y espalier. Bagged. | Drip | 1& 2 | 4000 | 5934 | 5771 |
| A Victoria. Y espalier. No bagged | Drip | 2 | 3500 | | 7931 |
| A Superior. Y espalier. No bagged | Drip | 2 | 3500 | | 10866 |
| A Marginal management | | 1& 2 | 0 | -720 | -720 |
| M Napoleon. Wood trellis | Flood | 1& 2 | 5100 | 4433 | 4509 |
| M Superior. Wood trellis | Flood | 1& 2 | 5100 | 7224 | 7014 |
| M Italia. Wood trellis | Flood | 1& 2 | 5100 | 3675 | 3679 |
| M Dominga. Wood trellis | Flood | 1& 2 | 5100 | 6995 | 6959 |
| M Red Globe. Wood trellis | Drip | 1 | 4620 | 7112 | |
| M Superior. Wood trellis | Drip | 1 | 3990 | 8754 | |
| M Superior. Galvan.-iron trellis | Drip | 2 | 3990 | | 7939 |
| M Red Globe. Galvan.-iron trellis. Mesh cover | Drip | 2 | 4620 | | 6163 |
| M Superior. Galvan.-iron trellis. Mesh cover | Drip | 2 | 3990 | | 13573 |
| M Superior. Galvan.-iron trellis. Mesh & plastic | Drip | 2 | 4550 | | 15635 |
| M Crimson. Galvan. -iron-trellis. Mesh cover | Drip | 2 | 4860 | | 17590 |
| M Marginal management | | 1& 2 | 0 | -787 | -787 |

(1) Areas: Alicante (A), Murcia (M). (2) $Q_i$ is annual water supply for i activity. (3) Net margins excluding manual labour cost. Source: Own elaboration

Table 3. Decisional variables of Spanish table-grape: description, net margin for scenarios and annual water supply

Water shadow prices are higher in Alicante, where smaller water allocations than in Murcia are allowed, and water is an even more valuable resource; thus, productivity of each cubic meter is higher even when technology is adopted. In any case, the resulting values are very high and caution is recommended when considering them. To determine what the real affordable price is, demand functions were calculated in such a way that the cost of the required water is subtracted from each resulting cropping plan. The cost of water is concordant with the price included in each parameterization. The results of this calculation are Figure 3, where it is observed that the production units in Murcia demand more water than those of Alicante, up to 0.60 €/m³, a price at which Murcia would begin to reduce consumption. The curves of scenario-2 show a higher demand than those of scenario-1 since, if the grower has technology, other limitations, such as labour, can be compensated, and more productive varieties will be planted, but these consume more water.

Since the demand for water only begins to decrease when prices are very high, availability of the resource is an even greater limitation than its price. Analyzing the repercussion of the price of water on net profit (Figure 4), it can be observed that with prices above 1.5 €/m³ only a very low profit is obtained, or there may even be losses. If a reference income is set at 18,000 euros to compensate the yearly work of the entrepreneur, the maximum price that small grape growers in Alicante can afford is 0.25 €/m³ in scenario-1 and 0.60 €/m³ in scenario-2. In Murcia, this reference income is achieved when prices are lower than 0.15 Euros per cubic meter in scenario-1, while the current price of water is 0.18 €/m³, meaning that only those growers with more technology are achieving profits. In the case of implementing the improvements of scenario-2, in Murcia it is possible to surpass the reference income when the price of water is not more than 1.1 €/m³.

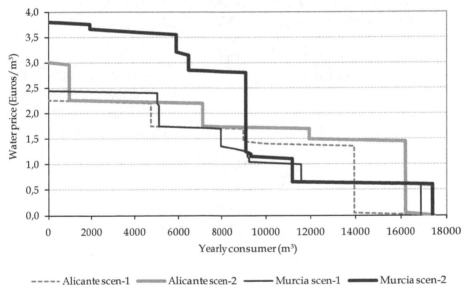

----- Alicante scen-1 ▬▬▬ Alicante scen-2 ▬▬▬ Murcia scen-1 ▬▬▬ Murcia scen-2

Fig. 3. Irrigation water demand curves in Spanish grape-table (Data for 5 ha production unit)

The main conclusions are:

-   To guarantee the continuity of the Spanish table-grape farms (most of them being small and family-run), the adoption of technological improvements seems to be essential. An essential improvement (being massively applied in both areas) is the increase of surfaces with drip-irrigation. This technique enables choice of the watering times for the plots, achieving an optimization of the allotments awarded in the plantation, these usually being lower than the theoretical needs of the crops.

- The scarce response of the demand, because of rising water prices, and therefore, of implementation of a pricing policy, denotes that the problem of these producing regions is its availability rather than its price. This conclusion is only valid if the growers have an acceptable profitability, which cannot be assured in the medium or long-term with the current market situation. To achieve the minimum profit fixed in this study, the price of water should not exceed 0.15 €/m³ in Murcia and 0.25 €/m³ in Alicante. These prices could be considerably higher if growers improve their level of technology. Therefore, the adoption of technology will be the most direct strategy for increasing expectations of continuing production of the growers, who, in general, do not feel capable of overcoming the iron rules of the markets.

| Proportion of activities (in %)[1] | Irrigation Type | Scenario-1 | | Scenario-2 | |
|---|---|---|---|---|---|
| | | $L_1$ | $L_\infty$ | $L_1$ | $L_\infty$ |
| ALICANTE (VINALOPO): | | | | | |
| Aledo. Traditional espalier. Bagged | Flood | | | 16.8 | |
| Aledo. Traditional espalier. Bagged | Drip | 40.5 | 40.5 | 23.7 | 40.5 |
| Aledo. Y espalier. Bagged | Drip | 4.5 | 4.5 | 4.5 | 4.5 |
| Italia. Traditional espalier. No bagged | Flood | 20.3 | 20.2 | | |
| Italia. Traditional espalier. No Bagged | Drip | 4.7 | 4.8 | 10 | 12.5 |
| Victoria. Y espalier. No bagged | Drip | | | 10* | 10* |
| Superior. Y espalier. No bagged | Drip | | | 5* | 5* |
| Proportion with marginal management | | 30* | 30* | 30* | 27.5 |
| **Total Net Margin (€)** | | **21,175** | **21,177** | **23,785** | **24,874** |
| **Total manual labour (h)** | | **1,345** | **1,345** | **1,210** | **1,254** |
| **Total water consumption (m³)** | | **13,898.5** | **13,899** | **13,841** | **14,428** |
| **Shadow irrigation prices (€/m³)** | | **4** | | **1,9** | |
| MURCIA : | | | | | |
| Napoleón. Wood trellis | Flood | 25 | 27 | 5 | 5 |
| Dominga. Wood trellis | Flood | 5 | 5 | 10 | 10 |
| Superior. Wood trellis | Flood | 10 | 8 | | |
| Superior. Wood trellis | Drip | 30 | 32 | | |
| Superior. Galvan. iron trellis. Mesh cover | Drip | | | 45* | 45* |
| Crimson. Galvan. iron trellis. Mesh cover | Drip | | | 10* | 10* |
| Proportion with marginal management | | 30* | 28 | 30* | 30* |
| **Total Net Margin (€)** | | **17,511** | **18,034** | **35,070** | **35,103** |
| **Total manual labour (h)** | | **1,835** | **1,873** | **2,154** | **2,158** |
| **Total water consumption (m³)** | | **16,185** | **16,607** | **15,233** | **15,285** |
| **Shadow irrigation prices (€/m³)** | | **1,24** | | **0,63** | |

1) Respect total surface (5 ha). * Limit coincident with the restrictions of the models.
Source: Own calculation

Table 4. Cropping plan and results for two scenarios in Spanish table-grape

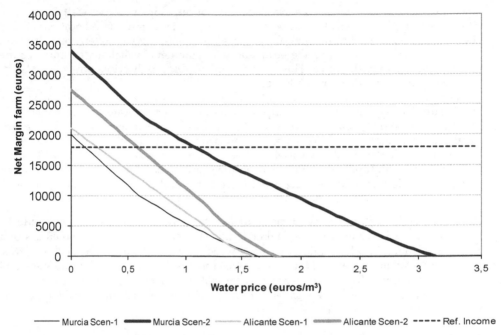

Fig. 4. Repercussion of the price of water on net profit in Spanish grape-table (Data for 5 ha production unit)

## 5. Conclusion

Water demand curves have been used to analyze the trends of the most important Mediterranean woody crops when up against tariff policies that differ from the current ones. These calculations have been applied to olives, almonds and vineyards grown in the inland regions of the Valencian Community, and table grapes in the two main production areas in the southeast Spanish Mediterranean region. The main conclusions derived from the two case studies are:

- The shadow price of water is very high for all crops under study, increasing in those scenarios with a low technological level. Technology can offset other agronomic technical limitations of the farm-holding, optimizing the production process, and therefore the shadow price of water is reduced. Usually farmers believe that technology is their best strategy to improve farm viability, although, on the other hand, technological improvements allow more productive varieties to be grown, which typically require more water, and as a result the more technologically developed scenarios are also more demanding of water resources.
- Irrigated crops with moderate irrigation requirements, such as table grapes, show highly inelastic demand curves, at least in the first price phases. As this resource becomes more expensive, demand falls while the surface area with marginal management increases, which may be the step prior to future crop abandonment. For the typical rainfed crops, the demand curves display strong inelasticity, demonstrating the huge value of having an extra cubic meter of water, this to be applied at specific

moments and in low doses. In general, Mediterranean crops make very efficient use of the water supplied, even at doses below agronomic irrigation needs. Allocations should be supplied at the point in time most crucial to the crop, which often coincides with reduced water availability on the farm, and in this moment availability becomes a more limiting factor than the price. This does not mean water high prices can be paid, since the real affordable price is much lower.

-   Water prices to be paid can theoretically be very high, as with respect to woody species their survival depends on timely supply, which does not indicate that sustained high prices are to be met. In fact, the first step taken by farmers is to increase the surface area with marginal management (prior step to abandonment) and concentrate investment in the more profitable fields or varieties. Therefore, it is foreseeable that a tariff policy implementing high prices would result in the gradual abandonment of Mediterranean crop cultivation and thereby reduce the economic activity in large tracts of land, especially in the inland regions.

Further research based on woody crops in areas where these are implemented would be desirable for the achievement of a sustainable and efficient use of water. The effects of a water pricing policy in other highly extended woody Mediterranean crops, such as citrus, almond, pomegranate or peach, is still unknown. This research would inform farmers and policy makers about reliability of water pricing policies in this kind of crops, avoiding undesirable effects on farmers and environment, and enforcing the reliability of the measures proposed by the WFD.

## 6. Acknowledgment

We are grateful to the financial aid received from the Spanish Ministry of Science and Innovation and the ERDF through the GEAMED project "*Gestión y eficiencia del Uso Sostenible del agua de Riego en la Cuenca mediterránea*" (AGL2010-22221-C02-01). Thanks to Dr. Pedro Caballero, for his comments on an earlier draft of this chapter.

## 7. References

Amador, F., Sumpsi, J.M. & Romero, C. (1998) A non-interactive methodology to assess the farmers' utility function: An application to large farms in Andalusia. Spain. *European Review of Agricultural Economics*, Vol. 25, pp. 95-109. ISSN: 1464-3618

Bazzani, G.M., Di-Pasquale, S., Gallerani, V. & Viaggi, D. (2004). Irrigated agriculture in Italy and water regulation under the European Union water framework directive. *Water Resource Research*, Vol. 40(7), W07S04. ISSN 0043-1397

Ballestero, E. & Romero, C. (1991). A theorem connecting utility function optimization and compromise programming. *Operational Research Letter*, Vol. 10, pp. 412-427. ISSN: 0167-6377

Ballestero, E., Alarcón, S. & García-Bernabeu, A. (2002). Establishing politically feasible water markets: a multi-criteria approach. *Journal of Environmental Management*, Vol. 65(4), pp. 411-429. ISSN: 0301-4797

Berbel, J. & Gómez-Limón, J.A. (2000). The impact of water-pricing policy in Spain: an analysis of three irrigated areas. *Agricultural Water Management*, Vol. 43, pp. 219-238. ISSN: 0378-3774

Berbel, J., Mesa-Jurado, M.A. & Pistón, J.M. (2011). Value of irrigation water in Guadalquivir Basin (Spain) by the Residual Value Method. *Water Resources Management*, Vol. 25, pp. 1565-1579. ISSN: 0920-4741

CAPA (Conselleria de Agricultura, Pesca y Alimentación, Generalitat Valenciana). (March 2011). Publicaciones. Informes del Sector Agrario y Boletines de Información Agraria. Available from http://www.agricultura.gva.es/web/web/guest/la-conselleria/publicaciones/boletin-de-informacion-agraria.

CARM (Comunidad Autónoma de la Región de Murcia). (March 2011). Estadística agraria regional. Superficie regionales. Series históricas regionales. Available from http://www.carm.es/web/pagina?IDCONTENIDO=1391&IDTIPO=100&RASTRO=c488$m1174,1390

Dyer, J.S. (1977). *On the relationship between goal programming and multiattribute utility theory*. Discussion paper 69. Ed. Management Study Center. University of California.

Easter, K.W. and Liu, Y. (2005). Cost Recovery and Water Pricing for Irrigation and Drainage Projects. World Bank Technical Paper, Washington, D.C., 55p.

Edwards, W. (1977). Use of multiattribute utility measurement for social decision making. In: *Conflicting objetives in decision* (Bell D.E.; Keeney R.L. & Raiffa H., Ed.). John Wiley & Sons, New York. pp. 247-276

Fernández-Zamudio, M.A. & De-Miguel, M.D. (2006). Sustainable management for woody crops in Mediterranean dry-lands. *Spanish Journal of Agricultural Research*, Vol. 4(2), pp. 111-123. ISSN: 1695-971-X

Fernández-Zamudio, M.A., Alcón, F. & De-Miguel, M.D. (2007). Irrigation water pricing policy and its effects on sustainability of table grape production in Spain. *Agrociencia*, Vol. 41(7), pp. 805-845. ISSN: 1405-3195

Fernández-Zamudio, M.A., Alcón, F. & Caballero, P. (2008). Economic sustainability of the Spanish table grape in different water and technological contexts. *New Medit*, Vol. VII, num. 4, pp. 29-35. ISSN: 1594-5685

Florencio-Cruz, V., Valdivia-Alcalá, R. & Scott, C.A. (2002). Water productivity in the alto río Lerma (011) irrigation district. *Agrociencia*, Vol. 36, pp. 483-493. ISSN: 1405-3195

Gómez-Limón, J.A. & Berbel, J. (2000) Multi Criteria Analysis of Derived Water Demand Functions: a Spanish Case Study. *Agricultural Systems*, Vol. 63, pp. 49-72. ISSN: 0308-521X

Gómez-Limón, J.A., Arriaza, M. & Berbel, J. (2002). Conflicting implementation of agricultural and water policies in irrigated areas in the EU. *Journal of Agricultural Economics*, Vol. 53(2), pp. 259-281. DOI: 10.1111/j.1477-9552.2011.00294.x

Gómez-Limón, J.A., Riesgo, L. & Arriaza, M. (2004). Multi-criteria analysis of input use in agriculture. *Journal of Agricultural Economics*, Vol. 55(3), pp. 541-564. DOI: 10.1111/j.1477-9552.2004.tb00114.x

Gómez-Limón, J.A. & Riesgo, L. (2004a). Water pricing: Analysis of differential impacts on heterogeneous farmers. *Water Resources Research*, Vol. 40, W07S05. ISSN 0043-1397

Gómez-Limón, J.A. & Riesgo, L. (2004b). Irrigation water pricing: differential impacts on irrigated farms. *Agricultural Economics*, Vol. 31(1), pp. 47-66. ISSN: 1574-0862

Gurovich, L.A. (2002). Irrigation scheduling of table grapes under drip irrigation: an approach for saving irrigation water and energy cost in Chile. *International Water Irrigation*, Vol. 22(2), pp. 44-50.

Huirne, R.B.M. & Hardaker, J.B. (1998). A multi-attribute model to optimise sow replacement decisions. *European Review of Agricultural Economics*, Vol. 25, pp. 488-205. ISSN: 1464-3618

Hearne, R.R. & Easter, K.W. (1997). The economic and financial gains from water markets in Chile. *Agricultural Economics*, Vol. 15(3), pp. 187-199. ISSN: 1574-0862

INE (Instituto Nacional de Estadística). (March, 2011). Estadísticas agrarias. Encuesta sobre la estructura de las explotaciones agrícolas. Available from http://www.ine.es/inebase/menu6_agr.htm

Jorge, R.F., Costa-Freitas, M.B., Seabra; M.L. & Ventura, M.R. (2003). Droughts: will farmers change therir decisions?. *New Medit*, Vol. 2(4), pp. 46-50. ISSN: 1594-5685

Keeney, R.L. & Raiffa, H. (1976). *Decisions with multiple objectives: preferences and value trade offs*. John Wiley & Sons, New York.

MARM (Ministry of Agriculture, Fisheries and Food of Spain). (March, 2011). Anuario de estadística. Available from http://www.marm.es/es/estadistica/temas/anuario-de-estadistica/default.aspx

Mesa-Jurado, M.A., Berbel, J. & Orgaz, F. (2010). Estimating marginal value of water for irrigated olive grove with the production function method. *Spanish Journal of Agricultural Research*, Vol. 8, special issue (S2), pp. 197-206. ISSN: 1695-971-X

Romero, C. & Rehman, T. (2003). *Multiple criteria analysis for agriculture decision*. Elsevier, Amsterdam. ISBN 0444503439

Sabuni, M. & Bakshoudeh, M. (2004). Determining the relationship between the opportunity cost of water and farmers'risk attitudes, using multi-objective programming. *Agr Sci Tec*, 18(1), pp. 39-47.

Sumpsi, J.M., Amador, F. & Romero, C. (1996). On farmer's objectives: A multi-criteria approach. *European Journal of Operational Research*, Vol. 96, pp. 64-71. ISSN: 0377-2217

Sumpsi, J.M., Garrido, A., Blanco, M., Varela, C. & Iglesias, E. (1998). *Economía y Política de Gestión del Agua en la Agricultura. Ministerio de Agricultura, Pesca y Alimentación*, Ed. Mundi Prensa, 84-7114-781-5, Madrid, Spain.

Varela-Ortega, C., Sumpsi, J. M., Garrido, A., Blanco, M. & Iglesias, E. (1998). Water pricing policies, public decision-making and farmers response: Implications for water policy. *Agricultural Economics*, Vol. 19, pp. 193– 202. ISSN: 1574-0862

Yu, P.L. (1973). A class of solutions for groups decision problems. *Management Science*, Vol. 19, pp. 936-946. DOI: 10.1287/mnsc.19.8.936

Zeleny, M. (1973). *Compromise programming. Multiple criteria decision making*. (Zeleny M., and Cochrane J., Ed.). University of South Carolina Press, Columbia SC. pp.262-301.

# 5

# Water Rights Allocation, Management and Trading in an Irrigation District - A Case Study of Northwestern China

Hang Zheng[1], Zhongjing Wang[1], Roger Calow[2] and Yongping Wei[3]
[1]State Key Laboratory of Hydro-Science and Engineering, Department of Hydraulic and Hydropower Engineering Tsinghua University, Beijing
[2]Overseas Development Institute, London
[3]Australia-China Water Resource Research Center,
Department of Civil and Environmental Engineering,
University of Melbourne, Melbourne, Victoria
[1]China
[2]UK
[3]Australia

## 1. Introduction

Demographic change, growing urbanization, intensification of agriculture and climate change all pose a continual challenge to the availability of water resources. The increasing competition for water demand among the sectors of human activities and for the environment requires the development of policies for water resources sustainability.

Policies to expand water resource supplies are currently not in vogue because they involve the regulation of water through physical impediments such as the construction of dams, weirs and channels. Over the last few decades demand management policies involving water pricing, assigning water rights and introducing water markets have received increased emphasis. Water rights, a prerequisite for water markets, are considered as a key water management instrument to improve water use efficiency.

In response to concerns of increasing water scarcity and seriously degraded river ecosystems, water policy in China over recent decades has shifted from investing in large storage and delivery infrastructure to policies and institutions designed to allocate the existing resource more efficiently. The definition and establishment of water rights allocation systems are important components of water management reform. Water rights allocation systems did not exist in China before 1988. The 1988 Water Law and its revision, the 2002 Water Law, have introduced initial water rights allocation across the country. In China, water rights are defined by the state according to the priorities assigned to competing users. Water resources in a trans-provincial (or prefectural) basin are shared amongst the jurisdictions administratively.

Northwestern China faces more severe water shortages for its arid climate. The agriculture water use is above 80% of total water use in this region. Therefore, agriculture water rights reform raises much concern currently. In some areas, the water rights defined for province or prefecture are allocated further to the irrigation districts and farmers. Then, the water trading happens in these places. For example, Hangjin Irrigation District on the south bank of the Yellow River, Inner Mongolia has traded some of its irrigation water to downstream factories. The trading is termed "irrigation water-saving supported by industrial investment, with saved water traded to industry". At the same time, Hangjin Irrigation District has conducted a comprehensive reform of irrigation water management focused on water rights.

This proposed chapter aims at introducing a framework for water rights allocation, management and trading in the farmers' lever, in order to address: (1) how the long-term water rights can be defined for the individual farmers in order to share the total water resource of the irrigation district; (2) how the farmers' water rights are administrated, monitored and accounted; (3) how the farmers to trade their water rights with the industry users or other farmers in the context of current Chinese Water Law.

The chapter will describe the current status of water management in the Hangjin district, outlines some of the problems water trading has produced, and presents a framework for further water rights reform focused on rights allocation, the granting of volumetrically-capped water certificates and tickets, water use planning and monitoring, and the responsibilities of water user associations in ensuring that individual farmers receive fair allocations. In additional, a water trading approach based on "water extraction period exchange" in Taolai irrigation distract, Gansu, China will be discussed in the chapter. The chapter then summarizes key recommendations of relevance to Hangjin and Taolai and other irrigation districts in China.

## 2. Water rights allocation among the farmers

### 2.1 Introduction

The Inner Mongolia Autonomous Region in China enjoys exceptional advantages. In particular, the region has an abundance of natural resources for the development of mining, electric power, metallurgy, chemical, and machinery processing industries. The Region plans to use these resources to build a large energy base in the "golden triangle" of Hohhot, Baotou and Ordos (Figure 2.1) to create an affluent society. However, the serious shortage of water resources hinders the development of the regional energy industry, and the region's allocation of water from the Yellow River Conservancy Commission (YRCC) is already fully committed. It is under such circumstances that that the autonomous region initiated a pilot program involving the transfer of water rights. Since 2003, a number of pilot projects for water right transfer have been launched by the YRCC and the Inner Mongolia Department of Water Resources (Shen et al. 2006), aimed at meeting the growing water needs of downstream industrial users.

One of the first such pilots has involved Hangjin Irrigation District. Beginning in 2004, the newly established Office of Water Rights and Transfer in Ordos city has overseen a program in which water saved through canal lining in the district is transferred to downstream industries, with the costs of lining met directly by the industrial beneficiaries. According to the Inner Mongolia Autonomous Region Water Rights Transfer Planning Report, in the three-year period from 2005 to 2007, 13 enterprises invested a total of RMB 600 million in canal lining. According to the plan, the implementation of the project will save as much as 138 million m3 of water. Industrial users funding the capital costs of canal lining are also obliged to meet the ongoing operations and maintenance costs of canal repair over a 25 year term.

The channel lining and water transfer program in Hangjin highlights one response to a wider problem in China – the problem of increasing scarcity and growing competition for water between uses and users. In this context, agriculture is under growing pressure to release water to urban and industrial users. Clear rules are needed for doing this and, increasingly, clear rights will be needed within irrigation districts (IDs) so that farmers can be confident about how much water they will get, and when they will get it. Moreover, a system of clearly defined, secure water rights provides the foundation for many other reforms aimed at managing demand and increasing efficiency, including water pricing and water trading.

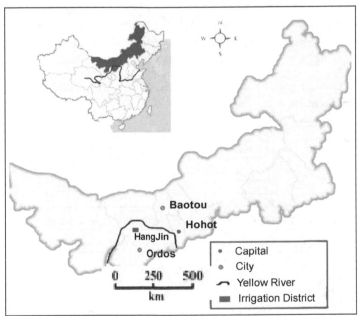

Fig. 2.1 Map of Inner Mongolia showing the Yellow River, major cities and Hangjin ID

## 2.2 Hangjin irrigation district

Hangjin County is located to the northwest of Ordos City in Inner Mongolia (Figure 2.1). Along its northern margin the Yellow River winds down with a length of roughly 253 km, making Hangjin County the longest flowing section of Yellow River of all counties nationwide. The county includes nearly 40,000 ha of designated farmland along the Yellow River, and is one of three major irrigation zones of Inner Mongolia. It is also one of China's main grain producing areas. Hangjin Gravity Irrigation District (HID) in Hangjin County – the focus of this study – is the only irrigation district in Ordos with the right to take water from the Yellow River. HID is located on the south bank of the Yellow River and covers an area of approximately 23,000 ha. Of this, roughly 21,000 ha is gravity fed and 1700 ha is pumped (at the head of the system).

Hangjin Irrigation District draws all of its water from the Yellow River. Its water use is therefore controlled, ultimately, by the YRCC, which sets minimum flow requirements for the river at provincial/regional boundaries based on an Annual Allocation Plan (Table 2.1), and allocates relative shares to individual provinces and regions according to supply and demand conditions. In a normal year, Inner Mongolia therefore receives 5.86 billion m³ out of a total flow of 37 billion m³. The maximum (sometimes termed 'normal') gross diversion to the Hangjin district –the permitted volume – is 410 million m³ per year, including a mandatory return flow of 35 million m³ per year. So, the normal net diversion to HID is 375 million m³. Return flows are fed back to the river through four main drainage channels. Savings of 130 million m³ per year from canal lining, traded out of the irrigation district, will leave an ongoing diversion of 280 million m³ per year, illustrated in Figure 2.2.

| Province/ region | Qinghai | Sichuan | Gansu | Ningxia | Inner Mongolia | Shaanxi | Shanxi | Henan | Shandong | Heibei & Tianjin | Total |
|---|---|---|---|---|---|---|---|---|---|---|---|
| Annual water use billion m³ | 1.41 | 0.04 | 3.04 | 4 | 5.86 | 3.8 | 4.31 | 5.54 | 7 | 2 | 37 |
| % | 3.8 | 0.1 | 8.2 | 10.8 | 15.8 | 10.3 | 11.6 | 15.0 | 18.9 | 5.4 | 100 |

Table 2.1 Water allocation in the Yellow River (YRCC, 2005)

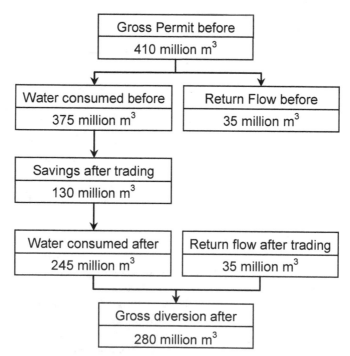

Fig. 2.2 Diversion, consumption and return flows for Hangjin Irrigation District

By 30 September 2006, a total of six canal lining subprojects had been completed, each funded by a separate industrial enterprise. The idea of "Industrial Investment in Water Saving for the Transfer of Agricultural Water Rights" has helped alleviate the water shortages experienced by industry, and has also helped reduce the burden of farmers by saving water and reducing farm costs. Currently, the annual water fee for each householder has been reduced by around 20-30 RMB/year. Farmers' costs have reduced because they no longer have to pay for water losses in the channels that deliver water to the point where water user associations (WUAs) make bulk purchases on behalf of the farmers they represent.

The channel lining and transfer project has had many benefits. However, trading has also created a number of problems, particularly for the irrigation agency that is responsible for managing and maintaining irrigation infrastructure above WUA purchase points – Hangjin Irrigation Management Bureau (HIMB). Moreover, the rights of farmers within the district remain ambiguous.

A framework for a modern system of volumetrically defined water rights in HID has been developed (WET, 2007). It is proposed that this serves as a template for guiding reform in other IDs in China as competition for water increases, and agricultural users face growing pressure to account for their water and release 'surpluses' to urban and industrial users.

The sections below discuss rights definition, allocation and management issues within HID. The principal focus is on improving the distribution of water within an ID so that farmers receive secure, transparent and equitable allocations within the overall permitted allowance of the ID.

## 2.3 Long-term Initial water rights allocation

Drawing on field work conducted in HID, WET (2007) describes how the water diverted to the district under its irrigation permit is currently allocated through main and branch canals, and down to individual farm households. In common with many IDs in water-scare northern China, the allocation process combines bulk volumetric charging to farmer groups (increasingly WUAs) established on branch canals, with area-based charging for farmers. Water User Associations purchase pre-paid water tickets on behalf of farmers, and are responsible for (amongst other things) distributing water within their command areas and collecting fees.

WET describe how water allocation to WUAs could be improved according to the principles of fairness, efficiency and environmental sustainability, amplified below. They also describe how the water rights of WUAs could be volumetrically defined and capped through the issue of Group Water Entitlements (GWEs) at the point at which WUAs pay for bulk deliveries. Below this point, farmers would continue to pay for water on an area basis, as delivery and monitoring infrastructure in Hangjin, and most IDs in China, is not in place to monitor individual entitlements at the household level.

A volumetric cap on the water rights of WUAs needs to fully consider existing patterns of water use within and between WUAs, and the experience of farmers, WUA representatives and HIMB staff in administering present systems. Hence it is proposed that rights allocation follows existing practice by linking land and water rights. In other words, rights assigned would be directly linked to the (existing) irrigated areas of each WUA, and could not be negotiated upwards by a WUA seeking to expand its irrigated area or plant more water-intensive crops, for example. Hence one objective of defining and enforcing WUA-based GWEs would be to end the requirements approach to water use planning that currently prevails so that, in future, water savings rather than additional supply would be used to maintain or increase farm production and farmer incomes.

Different regions and different groups of people should enjoy equal rights to water for survival and development. Hence the allocation of rights should guarantee fairness between different management sections of an ID, different WUAs and different water users and, in particular, afford protection to those farmers with small land holdings. In defining and allocating rights, consideration should also be given to 'third party' impacts on (linked) environmental services and other downstream users, such as groundwater users dependent on return flows from the irrigation district. How can the GWEs of individual WUAs be calculated to account for these factors, and to account for channel losses incurred to the points in the system at which WUAs purchase water? WET (2007) describe the calculations involved. A water allocation model is used in the water rights allocation process. The farmers' irrigation land area and crop mix are considered in the model (Figure 2.3).

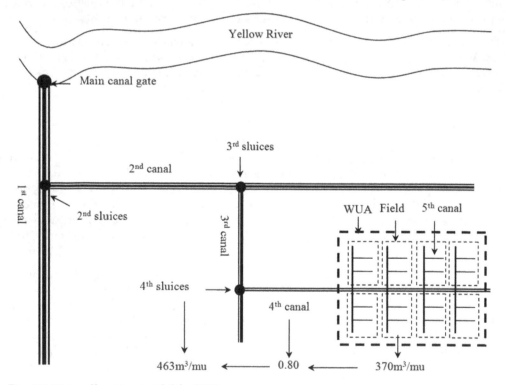

Fig. 2.3 Water allocation model for HID

The combined irrigated area of all 43 WUAs in the gravity flow section of HID is estimated at 21,322 ha. The total volume of water that needs to be delivered to fourth level sluices (and therefore WUAs), after subtracting losses in the canals above, is estimated at

143 million m³/year. The total volume of water that needs to be diverted from the Yellow River to meet WUA requirements and cover conveyance losses is 225 million m³/year. Total losses in the canals above fourth level sluices are estimated at 82 million m³/year. Using this data, and similar calculations covering allocations to individual WUAs, the long term, initial water rights of each WUA in HID can be determined as GWEs. These, in turn, form the basis for the issue of water certificates.

In contrast to the current farmer-driven approach to estimating water needs in ID, such an allocation provides a more scientifically-sound basis for defining and capping rights within the overall allowance of the irrigation district, and for accounting for all transmission losses through main and branch canals to WUAs. Since losses in each canal have now been estimated, future conservation efforts – including trading in transmission savings – can be better targeted and quantified. In this way, the approach to defining and allocating GWEs described above can form the basis for rights reform in other IDs.

## 3. Water rights management in an irrigation district

An integrated framework of irrigation water use in compliance with the farmers' water rights will be proposed in this section for Hangjin irrigation district, including the water use monitoring and accounting, accounting the farmers' water use to ensure that their water uses are under the allocation quota and water tickets as well as the role of water users association, et al.

### 3.1 Water rights certificates and water tickets

A system of water rights certificates can be used to formalise the rights of WUAs, providing information on long-term rights (defined by GWEs), annual water entitlements (defined by available supply in any given year) and the water purchased in each irrigation period. In addition, the system can provide information on any water transactions that have occurred between WUAs, and between WUAs and the irrigation management agency.  Table 3.1 provides a summary of certificate functions and uses.

| Function | Use |
|---|---|
| Voucher for long term rights | The irrigation management agency records each WUAs long-term water rights (GWEs) in a water certificate. |
| Calculation of purchase limits | At the beginning of the year, the irrigation management agency calculates the water purchase limit (annual entitlement) of each WUA and records this information on the certificate. After purchasing tickets in each irrigation period, the purchase amount will be recorded on the certificate to calculate the remaining purchase limit for the following periods. WUAs can purchase tickets up to the limit. |
| Record of water trading | The irrigation management agency records all information on water transactions. |
| Reference for water rights reallocation | The irrigation management agency will accumulate data on actual water use across seasons and between years, helping to guide any future adjustment. |

Table 3.1 Functions and uses of water certificates

To establish and operate such a system, the following steps are proposed (WET, 2007):

- After an initial water rights allocation process, the irrigation agency grants rights to each WUA in the form of a water certificate. This will show each WUAs long term water right.
- At the beginning of each year, the agency calculates the proportional water share that each WUA is entitled to (an annual entitlement) based on expected water availability in that year.
- Before each irrigation, the agency adjusts, as necessary, each WUAs annual entitlement in light of predicted supply to give a corresponding water purchase limit for all remaining irrigation periods. The purchase limit is recorded on each WUAs water certificate.
- After purchasing water tickets in any given irrigation period, the purchase amount is recorded on the certificate to calculate the remaining purchase allowance, or entitlement, of the WUA for the next period. In other words, a process of continuous water accounting is adopted between irrigation periods.
- Any water trading is recorded by the relevant agency section office on the water certificates of both buyer and seller. Trading with other sections is also checked and registered with the agency. Certificates would also show actual water deliveries after trading.

After a reasonable period of operation (5-10 years), the irrigation management agency can review certificates in light of actual water use and trading experience, and revise as necessary. Following any long term trade of water rights, the irrigation management agency can take back old certificates and issue new ones after thorough auditing and recording.

For each WUAs purchase of water, it is proposed that the current system of pre-payment through water tickets is continued. Water tickets provide the basis for water purchase, water delivery and water trading within prescribed limits. The ticketing system can ensure that both WUAs and the irrigation management agency have clear information on prices, deliveries and volumetric rights, allowing WUAs to trade savings freely (Wu & Wu, 1993). Water User Associations would buy water tickets according to their water certificates before each irrigation, and would also be allowed to purchase extra water from those WUAs deciding not to use their full allowance (Feng & Li, 1993). Table 3.2 provides a summary of ticket functions and uses.

| Function | Use |
|---|---|
| Support for permit control and quota management | WUAs buy tickets up to their caps; HIMB sells tickets according to water availability and water rights limits. |
| Pre-payment for water | Water is only supplied by HIMB once WUAs have purchased tickets. |
| Water trading and monitoring | WUAs can buy and sell 'saved' tickets; HIMB monitors ticket turnover and adjusts caps as necessary. |
| Payment voucher – rights and duties | Tickets provide information on GWEs, actual delivery and payment – a summary of entitlement and payment obligation. |

Table 3.2 Functions and uses of water tickets

In summary, water rights certificates would formalise the long-term water rights of WUAs within an ID. Water tickets would then 'translate' these rights into real-time rights for WUAs, allowing them to purchase water within the cap for a specific period, and according to how much water has been purchased previously. Long-term and real-time water rights are then connected through water use planning, which converts long-term GWE into the real-time water cap and water use scheduling according to the planned water demand and the runoff forecast of the river. The relationship between water rights, water rights certificates and water tickets is shown in Figure 3.1.

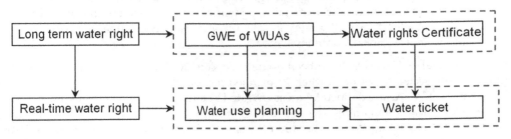

Fig. 3.1 Relationship between water rights certificates and water tickets

## 3.2 Water use planning

The objective of water use planning is to schedule water diversion, storage, delivery and use in an ID according to the requirements of farmers, available supply from the river and flow through the irrigation channel system. A water use plan is a guideline for the rational delivery and use of water within an ID, and can help improve irrigation efficiency and save water. In this section, it is proposed that the water use plan takes the GWEs discussed previously as a starting point, and then translates them into a real-time irrigation schedule for WUAs. WET (2007) propose that this occurs through a computer-based model that can balance demand and supply, guide allocation between WUAs and help manage rights in a quick and transparent manner.

At the beginning of the year, the annual water use plan for the ID would be prepared by the irrigation management agency, based on the annual water use plans submitted by each WUA (within capped limits), and submitted upwards through the irrigation agency to the higher level department for approval, such as the river basin management department. The river basin management department would then revise and approve the annual available water cap and the water scheduling of the ID, according to the water abstraction permit of the ID and the annual runoff forecast of the river. Afterwards, the irrigation district management agency would adjust the annual plan accordingly, and announce it to WUAs.

Prior to each irrigation, a WUA would then prepare and submit a plan for that period to the irrigation management agency for approval. The agency would check the available water allowance for each WUA, accounting for previous purchases, use under cap and overall irrigation scheduling, and make any necessary revisions or suggestions. Following ticket purchase, a final water use plan would be confirmed in accordance with sold ticket volumes and the scheduling needs of all WUAs.

The computer model would help managers prepare, modify, summarize and publish schedules, and could be interrogated quickly by all relevant stakeholders. The model would also help managers deal with the effects of runoff variation and hydrological uncertainty, including emergency planning in the event of floods or droughts.

### 3.3 Water users associations

A key element of irrigation reform is the promotion of WUAs as farmer run, participatory institutions that take the place of village leader-run water control organisations or government agencies, and take over management of water allocation and infrastructure management at a local level (Wang et al., 2006). Water User Associations are registered as legal entities under Chinese Company Law.

In HID, a total of 43 WUAs have been established since 2000 under 3rd level canals in the gravity flow sections, with a further 40 planned for completion by the end of 2008.. The boundaries of WUAs are defined by areas irrigated by tertiary and fourth canals. As a result, WUA and village boundaries do not always match. HIMB works with WUAs on the development of Annual Water Allocation Plans and scheduling arrangements, and WUAs are obliged to purchase water tickets prior to each irrigation period. It is proposed that WUAs hold and democratically manage GWEs on behalf of farmers and, within capped limits, continue to develop scheduling plans for household members, collect water fees, purchase water tickets from the ID management agency and undertake maintenance work on the infrastructure within their command areas.

The ability of WUAs in Hangjin (and elsewhere in China) to manage water rights effectively under capped GWEs depends on a number of different factors. WET (2007) identify four key pre-conditions, based on a survey of WUAs and farm households conducted in 2007.

Firstly, GWEs-based accounting through water certificates would need to be carefully monitored and enforced. The allocation system in HID combines bulk volumetric charging to WUAs established on branch canals, with area-based charging for farmers. Under such a system, the irrigation district management agency supplies water to WUAs on a contractual basis; contracts have no (current) legal authorization, but do specify the rights and obligations of both the agency and WUAs. Such contracts, or agreements, provide a type of group water right, albeit one of limited security. In Hangjin, moreover, the delivery of water to WUAs is governed by service contracts between WUAs and HIMB. Field work in HID (WET, 2007) suggests that these arrangements provide a sound basis for clarifying rights and responsibilities around water delivery and payment, and for the monitoring and recording of delivery and payment. They are recommended for other irrigation districts embarking on quota-based rights reform.

Secondly, infrastructure needs to be compatible with defined rights and local management capacity. Any discussion on water rights reform cannot be isolated from an understanding of the infrastructure that is available to deliver, monitor and record water flows. In Hangjin, and in most other IDs in China, irrigation systems have not been designed to deliver and record flows to individual farmers. In these circumstances, volumetric rights can only be defined, monitored and enforced down to the level of the WUA and, conceivably, to production teams managing tertiary canals. Hence in such systems it is proposed that capped rights are allocated to WUAs through GWE-based certificates, recognising that farmer-level entitlements cannot (yet) be implemented.

Thirdly, WUAs need well-specified management functions, authority and accountability. A key issue here is whether WUAs genuinely represent the interests of all farmers, and whether they have the capacity to resolve competing claims and disputes. In Hangjin and other IDs where WUAs have been established, the management functions and authority of WUAs are spelt out in a charter, or set of written rules. The ability of farmers to assert individual claims within the bulk GWE will therefore depend on whether WUAs act as genuine organs of democratic self-management, and whether elections required under their charter are held in an open, inclusive and fair way.  It is therefore suggested that the democratic management of WUAs is scrutinised closely by the ID management agency for a period of time after initial establishment. Periodic audits of WUA performance covering this and other tasks (e.g. financial book-keeping) are recommended.

Finally, WUAs require adequate resources. A common assumption in irrigation turnover programmes is that WUAs are better (than government agencies) at undertaking water allocation, distribution and fee collection in a cost-effective way. However, new obligations may be a serious burden on WUAs if they have been formed without adequate attention to their ongoing support needs. A key question in Hangjin and other IDs, therefore, is whether pressure to reduce government outlays – a key factor driving management transfer - has extended to an unwillingness to provide sufficient resources for WUAs to retain elected staff and carry out management tasks effectively, particularly in relation to long term water allocation, technical backstopping and maintenance.  It is therefore recommended that WUAs are allowed to retain enough ticket revenue to cover the salary costs of their full-time staff, and to cover operation and maintenance tasks within the WUA command area. Resourcing issues could be similarly monitored through periodic audit.

### 3.4 Water metering and monitoring systems

Many existing monitoring systems in China are crude, and need to be upgraded to support the operation and management of a modern water rights system. In HID, for example, water levels are measured using simple gauges, and flows are measured with traditional flow meters. All measurements are done by hand, with staff having to monitor and regulate flows through over 20 gates to WUAs. In a large ID this creates a very heavy workload for staff and at times of peak water demand, there may be a shortage of manpower.

Future pressure on IDs to release water for urban and industrial users may increase pressure for more accurate monitoring of allocations to WUAs. In this context, automated water monitoring systems may help solve current and future problems, saving labour and money and providing more accurate monitoring and regulation of increasingly scare water.

Design and use criteria a monitoring system needs to meet are outlined below (WET 2007).

- Automated monitoring and data transmission. Automated systems are more accurate and less-labour intensive than manual ones, eliminating the need for station staff to travel between and monitor individual sites.
- Rapid calculation and easy access to data. Data calculation and analysis should be quick and accurate, and data interrogation should be simple and direct. At present, data

enquiries in HID can only be answered by sifting through large numbers of paper records.

- Remote control and monitoring of main sluices. The irrigation management bureau should be able to operate sluices on the main and branch canals at least remotely, avoiding long distance travel for station staff and the need to spend many hours at individual sites.

- Transparency. It is important that an automated system retains the transparency of the existing system. In particular, WUA managers and farmers should have easy access to information on water deliveries to WUAs to build confidence in the quota-based certificate and ticketing arrangements.

- Affordability. Any upgraded system needs to be affordable in terms of both capital costs, and the ongoing costs of repair and maintenance. Benefits can help off-set costs, however, and are likely to include time (labour) savings for irrigation management agency, and water security-income gains for farmers (through more timely and reliable water delivery).

- Durability and security. An upgraded system must be able to cope with the sediment-laden inflows of the river, and not require constant adjustment and maintenance. It should also be equipped with alarms to increase security, and data security and virus protection should be included.

- Ease of use. Advanced systems must be capable of being operated and maintained by station staff.

## 3.5 An integrated framework for rights management in irrigation districts

Drawing on the discussion above, a broad water rights framework is proposed for HID and other IDs in China. The framework consists of three elements: institutions, irrigation services and regulations. These are described briefly below and illustrated in Figure 3.2.

The institutional component refers to the management institutions responsible for water allocation and delivery, including the relevant river basin management departments, ID management agencies and WUAs. The government river basin management department is responsible for allocating water and issuing water permits to IDs, and auditing their water use plans. No changes to existing allocation arrangements and responsibilities are proposed here.

Irrigation management agencies are mainly responsible for water allocation to WUAs. In this paper, it is proposed that they assume responsibility for the granting and overall management of water rights certificates and water tickets issued to WUAs, in addition to existing responsibilities for collecting water fees, preparing the water use plan of the irrigation district, and monitoring water deliveries to WUAs. Water User Associations, in turn, would assume responsibility for purchasing water tickets within the caps set by GWE calculations, and would manage and monitor allocations under the cap to individual farmers. Field investigations in Hangjin suggest that, where ticket-based payment and contracting systems are already established, the capped arrangements for allocating and purchasing water proposed in this section could be implemented fairly easily.

Irrigation services include the initial allocation of water rights, the issue of water certificates and tickets, water use planning, water delivery and operation of infrastructure. The permitted water abstraction volume of the whole irrigation district is allocated to WUAs through the initial water rights allocation process described, forming the basis for granting water rights certificates and the sale of water tickets. WUAs would purchase tickets within their allocated rights, prepare a water use plan and submit it to the irrigation district management agency for approval. The irrigation district management agency would then complete a water use plan for the whole district and issue delivery instructions to sluice operators, according to each WUAs water use plan and remaining ticket purchase allowance. Deliveries would be monitored and signed-off as they are now, with agency staff and WUA managers entering into seasonal contracts, and jointly monitoring and confirming allocations. The irrigation district management agency would record each WUAs available water, purchased water, and supplied water every year and every watering in their water rights certificates on a continual basis, in order to check the water account and guide water supply in the next period.

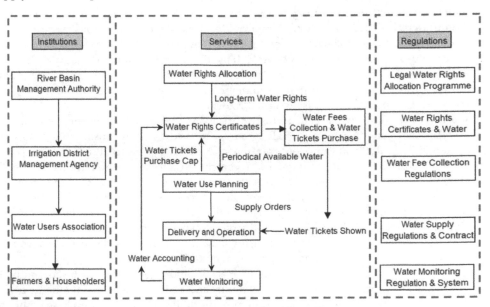

Fig. 3.2 A framework for water rights in an irrigation district

Regulations would then ensure effective implementation and monitoring of the services above, and would need to cover management regulations for the issue and use of water rights certificates and water tickets, water fee collection, water delivery and water monitoring. All management regulations and systems need to be carefully coordinated.

## 3.6 Recommendations

Based on field investigations in HID, a water rights framework for IDs in China has been proposed in this section, based on an initial water rights allocation, the issue of water rights certificates, sale of water tickets, water use planning and effective management of farmer-level rights through WUAs. Drawing on this framework, the authors offer the following recommendations:

1.  Group Water Entitlements should be defined and allocated to WUAs in HID and other IDs, and could additionally be given legal basis by government so that rights can be legally asserted and defended, providing greater security to WUAs and farmers. In addition, a water rights management system should be developed for all IDs, including regulations that cover water use planning, water delivery, emergency planning and risk management, the collection of water fees and maintenance of infrastructure. Entitlement-based allocation planning underpins future water conservation efforts and the development of a modern, socialist countryside of China.

2.  The use of an allocation plan to allocate water to WUAs in HID and other IDs is feasible. The annual allocation process in an ID needs to define and allocate GWEs within the overall permitted allowance of the district, determined by the relevant river basin authority. Allocation planning of this kind is fairer and more transparent than existing arrangements.

3.  Existing contract and ticketing procedures operating between HIMB and WUAs are well understood and respected. They provide an excellent platform for the introduction of GWEs and ticket-linked water certificates. Those WUAs that have set up systems of continuous water accounting between irrigations, and volumetric delivery to (and billing of) individual production teams, will be better able to meet new quota obligations in a fair and transparent manner. Such systems are recommended for other IDs in China embarking on rights-based reform.

4.  Water trading to downstream industrial users has reduced the revenue available to HIMB. The issue of funding will need to be addressed to ensure the long-term sustainability of the trading programme and channel infrastructure, and to protect farmers' long-term water rights. Management and institutional reforms in the ID should be conducted as soon as possible to improve management of the channels, enhance the financial position of the irrigation agency and secure new investment and financial resources. Most importantly, funding for the maintenance of newly lined channels in Hangjin should be secured from industrial enterprises as soon as possible. Similar channel lining and water transfer initiatives being considered by government agencies for other IDs in China need to learn from the experience of Hangjin.

5.  Information and monitoring systems in Hangjin and other IDs need to be gradually upgraded to improve accuracy and reliability and reduce manpower requirements. A key priority is to strengthen monitoring of water deliveries at WUA purchase points, as monitoring here affects both WUA payment and compliance with any new system of GWE-based water rights certificates.

## 4. Water trading among the irrigation districts under a duration-based water rights system

This section introduces a duration-based water allocation system, which has already existed for over 200 years in northwestern China, and discusses a water trading approach in the manners of exchanging the durations (the number of days) for water extraction. As case study in Taolai irrigation district, Gansu Province, China, the efficiency of the inner-agriculture water trading in the duration-based water allocation system is reviewed. This kind of water trading would provide possible approaches to promote water trading in Chinese irrigation district.

## 4.1 Introduction

Water resources support critical functions within human societies and ecosystems. Along with rapidly increasing population and improved living conditions, urbanization and industrial growth have led to increased demand, competition and conflicts between different water-use sectors (Liu et al., 2009). Climate change will intensify the situation in many parts of the world. It is very important to develop solution strategies to prepare against future conflicts.

The water rights system has been proved an effective tool for water resources management (Wang, 2009; Brook and Harris, 2008). Generally, water rights are defined in volumetric terms, with a statement of the probability that the nominal volume will be delivered in full in any given year (Productivity Commission, 2003). The predictability is a key requirement of a water rights system, so that users can have a reasonable expectation of the volume of water that will be available to them (Speed, 2009). In Australia, the water management authority announces an available percentage of the water rights volume to each stakeholder seasonally according to current reservoir level and inflows over the forthcoming season (Rebgetz et al, 2009). The announcement of the available water should be transparent and least variable to the stakeholders during the year, who thus take the minimum hydrological risk when using water. In the contrary, the water authority takes most of the responsibility for guaranteeing the water rights, which increases its management and technical cost. How to reduce the hydrological risk and to share it between the water manager and users in water allocation is still an ongoing issue, which raised a lot of studies recently both in the developed (Robertson, 2009; Zaman et al, 2009) and developing countries (Wang and Wei., 2006; Zhao et al., 2006; Hu and Tang, 2006; Zheng et al, 2010).

Some useful techniques and methods were proposed in these studies, including the long-term runoff predication, seasonal water allocation, self-adaptive water operation and so on. While all these techniques were developed to provide more reliable water volume availability under a centralized storage management, due to the hydrological uncertainties and storage capacity constraints, the hydrological risk affecting the volumetric water delivery cannot be completely removed only through these techniques. Moreover in practice, it is unlikely that dam managers will have complete information on user's water demand preferences. With this asymmetric information, a central manager may implement a sub-optimal release (allocation) policy, raising a problem that the intra-seasonal allocation is overly conservative, that is, where early season allocations are low and there is unallocated water available in storage (Hughes, 2009).

Institutional innovation such as redefining water entitlements rather than a share of total volume releases (natural stream runoff) is required. A system of allocating property rights to water from shared storages (as well as a share of inflows and losses), which is called capacity sharing, is established in Australia (Dudley and Musgrave 1988, Hughes, 2009). The capacity sharing proposed a decentralize the process allowing individual irrigators to exercise a degree of control over storage decisions and resulted in water entitlements more closely reflecting the physical realities of the water supply system: constrained storage capacity, variable water inflows and significant storage and delivery losses, and thus provided a solution to address the problems outlined above including hydrological risk and asymmetric information.

Similar with the capacity sharing, a Chinese traditional water entitlement may provide another way for solution. China has a long history of water resources development and management. Water diversions for irrigation dated as far back as 316 BC (Wouters et al., 2004). In 18th Century, the administrative water allocation appeared in some arid rivers northwest China by defining the order and length of water extraction period between upstream and downstream users. This kind of water allocation has been widely adopted in the northwest China for hundred years and is still used currently. This traditional water entitlement, instead of sharing the water extraction volume, allocated water rights based on water extraction duration. Each entitlement holder in the river basin is allocated a share of the total number of water extraction days. This water rights arrangement is named "duration-based water rights" in this paper.

The "duration-based water rights" is defined as a kind of water usufruct which is quantified by the independent duration of water extraction. In "duration-based water rights" system, the water users can store or withdraw the entire natural stream within their permitted extraction period, and manage it independently: determining how much water to use (or sell) and how much to leave in the water course, meanwhile taking all risks from the hydrological uncertainty and variation by themselves. The dam manager does not need to make volumetric allocation announcements and their role becomes in charge of water accounting: recording each user's inflows and withdrawals to monitor the quantity of water in each user's account. However, due to lack of volumetric cap in water use, the surface stream would be likely used out and the ground water would be over extracted in the "duration-base water rights" system.

## 4.2 Taolai River Basin

Taolai River Basin is an inland watershed located in northwest of China, covers an area of 28,100 km2. The total renewable water resources of the basin are estimated at 1.21 billion m3. It has three main water users: Jiuquan Iron & Steel Corporation (JQI&SC), Taolai Irrigation District (Taolai ID) and Yuanyang Irrigation District (Yuanyang ID) (Figures 4.1 and 4.2). The "duration-based water rights" started in Qing Dynasty about 200 years ago and is still used in this Basin. The stakeholders share the annual water extraction days (365 days in total) in the mainstream of Taolai River: 37 days of water use duration for upstream JQI&SC; 153 days for Taolai ID, and 175 days for downstream Yuanyang ID. These days named as "allocation durations" in this paper are shown as the horizontal length of the slices in Figure 4.3. The users are able to store or use the entire natural stream during their water allocation periods independently, as shown in the right vertical ordinate of Figure 4.3. However, due to lack of volumetric cap in water use in this "duration-based water rights" system, the water resources development ratio in Taolai River Basin is rising and close to 100% recently. An urgent institutional innovation is needed.

## 4.3 A water allocation-trading framework for duration-based water rights system

An improved "duration-based water rights" system is proposed by (1) introducing the volumetric water use cap in each allocation period, according to the water demand and historical water usage of the users; (2) creating the enabling environment for water trading; (3) promoting the water trading in the valley and (4) setting up an integrated water allocation-trading framework support these improvements (1), (2) and (3).

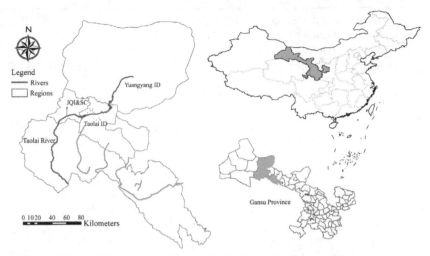

Note: Data from the National Fundamental Geographic Information System, China.

Fig. 4.1 Location of the Taolai River Basin, Gansu province, China

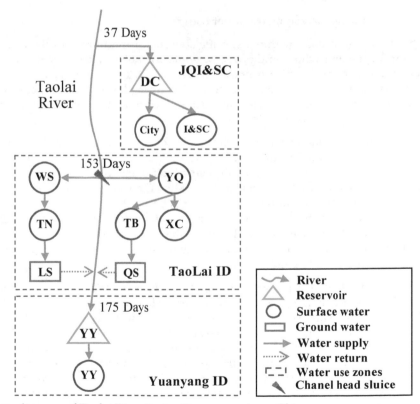

Fig. 4.2 Schematic diagram of Taolai River Basin, Gansu province, China

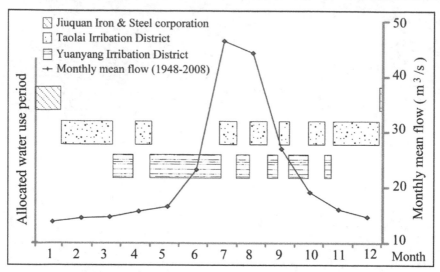

Fig. 4.3 Intra-year allocations of the "duration-based water rights" in Taolai River Basin, China

1.   Introducing the volumetric water use cap

The discharge volume within a specific allocation period provides the maximum available water for the user who is authorized to withdraw water in that period under the "duration-based water rights", which is shown in Figure 4.4. The shaded area under the flow curve indicates the available water volume for Yuanyang ID in its first allocation period of the year. The annual available water can be identified by accumulating all the available water in the allocation periods across the year.

The annual available water and historical water use of Taolai and Yuanyan ID under their "duration-based water rights" are shown in Figure 4.5 and Figure 4.6. The water volume is ranked descendingly by the total annual runoff of the Taolai River 1980-2008, and plotted versus the hydrology frequency of the years. The year with hydrology frequency of n% means that the annual runoff of the year will be exceeded in n years out of 100. The annual available water of the IDs is the accessible water within their allocation periods so that is part of the total annual runoff; therefore, the available water in a dry year may be larger than that in a wetter year for the inter-annual variability of the runoff process, which can be found in Figure 4.5 and 4.6.

In Figure 4.5, the historical water use of Taolai ID is stable and less than its available water, which indicates that there is some water didn't or can't be used by Taolai ID in its "duration-base water rights". Actually, this unused water is mainly made up of the flood in July and August, which can hardly be stored by Taolai ID without enough reservoirs in it, and was spill out to the ecosystem and downstream Yuanyang ID. For the ecological benefit from the flood water, involving the stream flow maintenance and groundwater recharge, the annual water use limit is introduced underneath the available water of Taolai ID and portrayed by the upper cap line of historical water use for satisfying the current water demand.

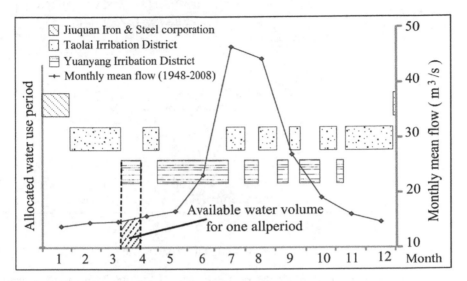

Fig. 4.4 Water available volume in the allocation periods of Taolai River Basin

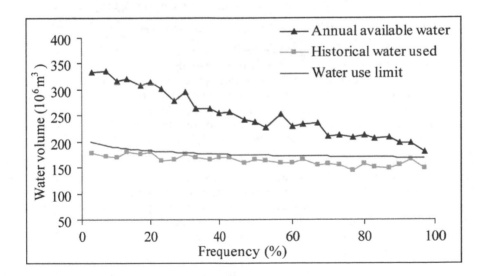

Note: Source from Zheng, 2011.

Fig. 4.5 Annual available water and water use limit for laolai ID

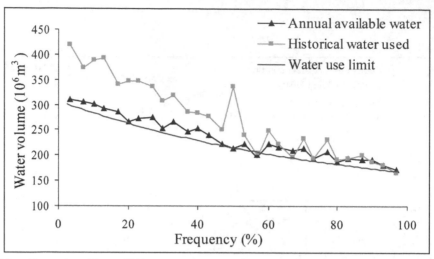

Note: Note: Source from Zheng, 2011. .

Fig. 4.6 Annual available water and water use limit for Yuangyang ID

In the downstream, Yuanyang ID stored and used part of the flood which was spilled by Taolai ID in its allocation periods, by a large reservoir in it. Therefore, the water used by Yuanyang ID was more than the water available in its allocation periods during 1980-2008. Water use limit of Yuanyang ID is established as the lower cap of its annual available water so that the Yuanyang ID's water extract can follow its "duration-based water rights" strictly and all the flood water can be released for ecosystem, as shown in Figure 4.6.

2.    Water use plan and information exchange for water shortage

The objective of water use planning is to schedule water diversion, storage, delivery and use in an ID according to the requirements of farmers, available supply from the river, and flow through the irrigation channel system. A water use plan is a guideline for the rational delivery and use of water within an ID, and can help improve irrigation efficiency and save water (Zheng and et al, 2009). It is proposed that the water use plan takes the water use limits discussed previously as a starting point, and then translates them into a periodical irrigation schedule for water users associations (WUAs) or farmers.

Prior to each irrigation (or allocation period), the period water use plan for the ID would be prepared by the Irrigation Management Agency, based on the plan submitted by each WUA, and submitted upwards through the irrigation agency to the higher-level department for approval, such as the River Basin Management Department, who would then revise and approve the water scheduling of the ID, in term of the water use limit of the ID. Afterwards, the Irrigation District Management Agency would adjust the plan accordingly, and announce it to WUAs and farmers.

If the irrigation water demand is not fully satisfied in the approved plan, on the agreement of the farmers, the Irrigation District Management Agency would release its water shortage information to the valley and search for the water sellers to promote a water trading. This process is proposed to occur through an on-line information exchange system that can balance demand and supply, guide pricing and help manage water trading in a quick and transparent manner, such as the "watermove" system in Australia (Available at https://www.watermove.com.au/Default.aspx).

## 3. Water trading

The predictability and transferability can be satisfied more strongly in the "duration-based water rights" system due to the stable water allocation periods and the decentralized management of the runoff within them. The economic efficiency of the water trading is described from Equation 4.1 to 4.4.

$$W_a = \int_0^{D_a} Q_a \cdot dt \tag{4.1}$$

$$W_j = \int_0^{D_j} Q_j \cdot dt \tag{4.2}$$

$$MU_a \succ MU_j \tag{4.3}$$

where, $W_a$ and $W_j$ describe the exchanged water volume in April and July between Taolai and Yuanyang IDs (m³); $Q_a$ and $Q_j$ are mean stream flow in the two months (m³/s); $D_a$ and $D_j$ are the number of exchanged days. $MU_a$ and $MU_j$ indicate the marginal utility of the water in April and July. From Figure 4.4, it is shown that $Q_a < Q_j$ and $D_a = D_j$. Therefore, $W_a < W_j$ which indicates Taolai ID obtained less water from Yuanyang ID in spring and gave more water back in summer. Due to the serious runoff insufficiency and irrigation competition in spring, the water is more valuable then, as shown in Equation 4.3. So, there is a possibility that the benefit gained by Taolai ID from the allocation period exchange in spring can be equal to its benefit loss in summer. If this balance happens (Equation 4.4), the water trading will be efficient.

$$MU_a \cdot W_a = MU_j \cdot W_j \tag{4.4}$$

In practice, water trading in the manner of exchanging water extraction days between upstream and downstream users has existed in the Taolai River Basin for years. This kind of water trading has being carried out in Taolai River Basin for yeas (totally 10 times, 2005-2009) and reallocated water effectively, with no need of seasonal water allocation, lower transaction cost and thus higher accessibility. In 2008, to solve the upstream water shortage caused by the mismatch between the irrigation schedule and allocation period distribution, the allocation period of Taolai ID was extended in April for 9 days, with the equivalent number of days reduction for Yuanyang ID simultaneously; while in summer when there is excess water for Taolai ID, the allocation period changed in the opposite directions as the same amount of days as in spring, shown in Figure 4.7.

## 4. Towards an integrated framework for water allocation-trading in the system

Drawing on the discussion above, broad water rights framework, combining the volumetric water use cap and the "time-based water rights", is proposed. The framework consists of three elements: institutions, water allocation-trading services and regulations. These are described briefly below and illustrated in Figure 4.8.

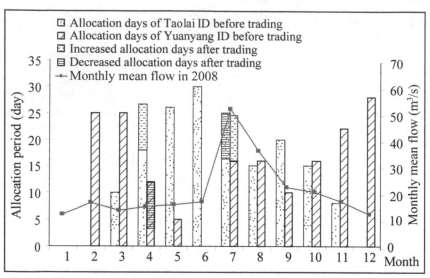

Fig. 4.7 Allocation period exchanges between Taolai ID and Yuanyang ID in 2008

The institutional component refers to the management institutions responsible for water allocation, trading and delivery, including the relevant State Water Resources Management Department, River Basin Management Authority, Irrigation District Management Agency and WUAs. The Water Resources Management Department of state government takes charge of administrative management for River Basin Management Authority. The authority is then responsible for allocating water and issuing water trading permits to IDs, and auditing their water use plans. Irrigation Management Agencies are mainly responsible for water allocation to WUAs and organizing a democratic decision making process for water trading. In this paper, it is proposed that they assume responsibility for preparing the water use plan of the irrigation district, and monitoring water deliveries to WUA. Water User Associations, in turn, would manage and monitor allocations under the cap to individual farmers.

Water allocation-trading services include issue the available water and water use limit, water use planning, water delivery and operation of infrastructure, as well as the information support, application approval, contrast and publicity for water trading. Prior the irrigation, the available water volume and water use limit for current allocation period would be issued to the irrigation district according to the duration of its allocation period and the forecasted runoff. Then, the rationing water volume of the whole irrigation district is allocated to WUAs through the normal volumetric water allocation process, providing the cap for water use planning of the WUAs. The Irrigation District Management Agency would then complete a water use plan of the whole district accordingly and check whether there are water shortage and the necessity for buying water. After the democratic consultation with farmers, if the irrigation district decides to buy some water and extend its allocation period, the management agency would publish its requirement to other irrigation districts and seek the water seller. If the buyer and seller get an agreement on water trading, they would submit a trading application to the River Basin Management Authority for approval.

The water trading will be legally effective only after the trading application is approved by the government and passed through by publics.

Regulations would then ensure effective implementation and monitoring of the services above, and would need to cover management regulations for the issue and use of volumetric water use cap together with the "duration-base water rights", and the information exchange, decision making, third-party impacts assessment and approval for water trading, as well as water delivery and water monitoring. All management regulations and systems need to be carefully coordinated.

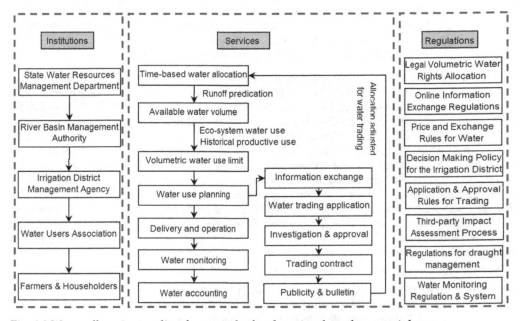

Fig. 4.8 Water allocation-trading frameworks for duration-based water rights system

## 4.4 Recommendations

A water allocation-trading framework based on the "duration-based water rights" was proposed in this section. Comparing with the normal volumetric water rights system, the framework is supposed to reduce the management cost of the water authority by introducing a decentralised and semi-independent water management to the stakeholders. In the framework, water entitlement is indicated by the fixed amount of water extraction days and would hardly be affected by the hydrological uncertain. The water users can manage the stream flow and storage independently during their allocation period under the volumetric water use cap. The water authority is just responsible for the water use planning and accounting, making sure the water use of the stakeholders not over their limit. The hydrological risk is shared mostly by water users while they get more flexibility to make their own storage decisions, taking into account their private information on water needs.

The idea of "duration-based water rights" is similar with the capacity sharing in Australia to a certain extent. Capacity sharing is a system of allocating property rights to water from shared storages proposed by Dudley (Dudley and Musgrave 1988, Dudley and Alaouze 1989, Dudley 1990, Dudley 1992). Each entitlement holder in an irrigation system is allocated a share of the total system storage capacity, as well as a share of total inflows (spill water and losses). Users are able to manage these capacity shares independently, as well as take the hydrological risk and losses. The duration-based water allocation-trading framework suggests a property rights system by sharing the water extraction duration, rather than storage capacity. But both of the systems suggest a decentralize the process by designing some system of property rights allowing individual irrigators to exercise a degree of control over storage decisions, which is helpful to address some of the problems of centralized water management, such as the hydrological risk, asymmetric information and transaction costs in water delivering. The proposed duration-base water allocation framework could provide a comparison reference for capacity sharing, and the success of the capacity sharing practice in Australia could be helpful to understand the feasibility and practicability of the proposed framework.

Volumetric water use limit was introduced and combined with duration-based water allocation in the study, which suggested a mechanism to integrate the international contemporary water rights system with the Chinese traditional water management. In recent decades, with the introduction of the global experiences of water rights reform, volumetric water right and its allocation system have been implemented across the China to a varying degree (Gao, 2006; Shen & Speed, 2009), and replaced the traditional water allocation system in most of the rivers. This has raised many conflicts in the reforms. The integrated mechanism proposed in this chapter would be helpful to buffer the conflicts when establishing a volumetric water rights system in valleys where the traditional water allocation is still working. Moreover, the integrated water allocation-trading framework could be used in the upcoming process of establishing the water market in Taolai River Basin, which would probably become the first water market in China and also a significant improvement in China's water rights reform. The framework would be feasible for the arid river, especially the valley which has uneven spatial distribution of the storage capacity.

For the limitation of the data and practices, the proposed framework just provides a conceptual framework of integrating the volumetric and time-based water rights without enough data verification. As noted, to transform this result based on one case study into the business of managing water catchments on a daily basis requires considerable further research, policy development and investment. Some future researches are still required to improve the framework, involving (1) defining the volumetric ground water cap in Taolai valley; (2) pilot study to verify the feasibility and validity of the framework; (3) modelling the irrigation water use planning which is constrained by the volumetric cap and time-based allocation in the irrigation districts and farmers level; (4) the technique for monitoring and accounting the water use and trading volume.

# 5. Acknowledgment

The study is supported by National Natural Science Foundation of China (51009076); International Science and Technology Cooperation Program of China (2010DFA21750); the project of "a framework for integrating the traditional water allocation and modern management" funded by water resources management department, Gaunsu Province, China; and the China Environment Development Program- Shiyang/Shule River Project as well as the 2011 Chinese Ministry of Water Resource Program - Shiyang River Project (201101046).

# 6. Reference

Babel, M. S.; Gupta, D. A. & Nayak, D. K. (2005) A model for optimal allocation of water to competing demands, *Water Resour. Management*, No.19, pp. 693–712.

Yellow River Conservancy Commission. (2005). Decree on Yellow River Water Resources Regulation Available from:
http://www.yellowriver.gov.cn/ziliao/zcfg/fagui/200612/t20061222_8784.htm
(21 December 2008).

Dudley, N. & Musgrave, W. (1988) .Capacity sharing of water reservoirs, *Water Resources Research*, No.24, pp. 649-658.

Dudley, N. (1990) *Alternative institutional arrangements for water supply probabilities and transfers.* Centre for Ecological Economics and Water Policy, University of New England, Armidale, NSW, Australia.

Dudley, N. & Alaouze, C. (1989) Capacity sharing and its implications for system reliability, Proceedings of the National workshop on Planning and Management of Water Resource systems, *Australian Water Resources Council Conference Series*, No. 17, 1989.

Dudley, N. (1992). Water allocation by markets, Common property and capacity sharing: Companions or Competitors? *Natural Resources Journal*, Vol.32, No. 4, pp: 757-778.

Feng, E. & Li, X. (2006). The study on the water rights system in the irrigation districts in the inland river basin, *China, Gansu Agricultural University Journal* [in Chinese] No.2, pp. 105–108.

Gao, E. (2006). *Water Rights System Development in China,* Beijing: China Water and Hydropower Publisher. [In Chinese]

Hayman, P.; Crean, J.; Mullen, J. & Parton, K. (2007). How do probabilistic seasonal climate forecasts compare with other innovations that Australian farmers are encouraged to adopt? *Australian Journal of Agricultural Research*, No.58, pp. 975-984.

Hu, J., & Tang, D. (2006). Study on the updating management of water right of Tarim River Basin. *Yangtze River*, No.37 (11), 73-75. [In Chinese]

Hughes, N. (2009) Management of irrigation water storages: carryover rights and capacity sharing, *Australian Agricultural and Resource Economics Society Conference Paper*, Cairns Queensland, 11-13 February.

Inner Mongolia Water Resources and Hydropower Survey & Design Institute (2005) *Inner Mongolia Autonomous Region Water Rights Transfer Planning Report* [in Chinese], Technical Report.

Liu, D.; Chen; X. & Lou, Z. (2009) A model for the optimal allocation of water resources in a saltwater intrusion area: a case study in pearl river delta in China, *Water Resour. Management*, No.24, pp.63-81.

Productivity Commission. (2003). Water rights arrangements in Australia and overseas. Commission Research Paper, Melbourne: Productivity Commission, pp: XIV.

Qureshi, M. ; Shi, T. & Qureshi, S. et al. (2009). Removing barriers to facilitate efficient water markets in the Murray-Darling Basin of Australia, *Agricultural Water Management*, NO.96, pp. 1641-1651.

Robert, B. & Edwyna, H. (2008). Efficiency gains from water markets: empirical analysis of Watermove in Australia, *Agricultural Water Management*, No.95, pp. 391-399.

Rebertson, D.; Wang QJ. & McAllister, A. et al. (2009). A Bayesian network approach to knowledge integration and representation of farm irrigation: 3 spatial application. *Water Resource Research*, No.45, doi:10.1029/2006WR005421.

Rebgetz, D.; Chiew S. & Malano H. (2007). Forecasts of seasonal irrigation allocations in the Goulburn Catchment, Victoria. The Modelling and Simulation Society of Australia and New Zealand Inc papers.

Shen, D. (2004). The 2002 Water Law: Its impacts on river basin management in China, *Water Policy*,No.6, pp. 345–364.

Shen, D.; Sheng, X.; Wang, R. & Liu, A. (2006) .Water rights reform in Inner Mongolia, China, *China Water Resource* [in Chinese], No.21, pp. 9–11.

Shen, D. & Speed, R. (2009) Water resources allocation in the People's Republic of China, *International Journal of Water Resources Development*, Vol. 25, No. 2, pp. 209–225.

Speed, R. (2009) A comparison of water rights systems in China and Australia, *International Journal of Water Resource Development*, Vol. 25, No. 2, pp. 389-405.

Wang, G. & Wei, J. (2006). *Water resources operations model in a river basin and its application.* Science Press, Beijing, China, pp. 109-112. [In Chinese]

Wang, X.; Huang, L. & Rozelle, S. (2006). Incentives to managers or participation of farmers in China's irrigation systems: which matters most for water savings, farmer income and poverty? *Agricultural Economics*, No.34, pp. 315-330.

Wang, ZJ.; Zheng, H. & Wang X. (2009) A harmonious water rights allocation model for Shiyang River Basin, Gansu Province, China, *International Journal of Water Resources Development*, Vol. 25, No. 2, pp. 355-370.

WET (2007) Water Entitlements and Trading Project (WET Phase 2) Final Report December 2007 [in English and Chinese] (Beijing: Ministry of Water Resources, People's Republic of China and Canberra: Department of the Environment, Water, Heritage and the Arts, Australian Government). Available from http://www.environment.gov.au/water/action/international/wet2.html.

Wouters, P.; Hu, D.; Zhang, J.; Tarlock, D. & Andrews-Speed, P. (2004). The new development of water law in China, *University of Denver Water Law Review*, Vol. 7, No.2, pp: 243–308.

Wu, D. & Wu, Z. (1993). On water use planning and water tickets, *Water Economy*, No.8, pp. 50-52. [in Chinese]

Xie, J. (2008). *Addressing China's Water Scarcity: A Synthesis of Recommendations for Selected Water Resource Management Issues* Washington, DC: World Bank Publications.

Zaman, A.; Malano. H. & Davidson B. (2009) An integrated water trading-allocation model, applied to a water market in Australia, *Agricultural Water Management*, No.96, pp. 149-159.

Zhang, C. & Wang, Z. (2004). The quantitative evaluation of human activities in Hei River Basin. *Earth Science*, No.29, pp. 396-390. [In Chinese]

Zheng, H.; Wang, ZJ.; Liang Y. & Roger C. (2009). A water rights constitution for Hangjin Irrigation District, Inner Mongolia, China, *International Journal of Water Resource Development*, Vol. 25, No. 2, pp. 371-385.

Zheng H.; Wang ZJ. & Hu S. (2010). A real time operation model for water rights management, *the 5th International Symposium on IWRM and 3rd International Symposium on Methodology in Hydrology*, Najing, China.

Zheng H.; Wang ZJ. & Hector M. et al. (2011). A water allocation-trading framework for duration-based water rights system in China, *Agriculture Water Management*, under review.

Zhao, Y.; Pei, Y.; & Yu, F. (2006). Real-time dispatch system for Heihe River basin water resourc. *Journal of Hydralulic Engineering*, Vol. 37, NO. 1, pp. 82-88. [In Chinese]

Zhou, W. (2001) The irrigation and water saving method in northwest China. *Advances in Water Resources and Hydropower*, Vol. 21, No. 1, pp. 1-4. [In Chinese]

# 6

# Irrigation Institutions of Bangladesh: Some Lessons

Nasima Tanveer Chowdhury
*Environmental Economics Unit (EEU)*
*Department of Economics*
*Gothenburg University, Gothenburg*
*Sweden*

## 1. Introduction

The objective of this chapter is to highlight some issues of existing irrigation institutions and their impact on cost and price of irrigation water in Bangladesh agriculture. Water is scarce in winter and agriculture is the major water using sector for irrigation in Bangladesh. The study mainly deals with how public institutions and water markets have evolved over time in response to changes in irrigation technology and how they affect the cost and price of irrigation water. There are many government run irrigation projects in the Northwest region (NW) of Bangladesh. Recently Bangladesh Water Development Board (BWDB) has handed over these irrigation projects to water user groups (mainly medium and large farmers) for management and cost recovery.

At present there are 5 types of irrigation systems which have given rise to different institutional setups. These institutions play a crucial role in water pricing. Both government and the private sector have been practising various methods over time for water allocation to the farmers' fields. The objective of this study is to examine how these institutions have come into existence as they are responsible for shaping water prices both in the public and private sector. Currently the public sector is responsible for maintaining the surface water irrigation projects which is only 10 percent of the total irrigation. The rest is the private sector mainly groundwater irrigation using minor irrigation devices. In the NW region irrigation is almost entirely dependent on groundwater due to scarcity of surface water. In December 2007 I had a focussed group discussion in the NW region of Bangladesh and use the results to understand how irrigation institutions are working in the region and how it affects irrigation cost and volumetric price per unit of water. Farmers in Bangladesh do not pay per m$^3$/litre for irrigation water; they pay for pumping cost if any and labour charges when required.

Bangladesh is an agricultural country divided into 7 hydrological regions. The average annual rainfall varies from 1,200 mm in the extreme west to over 5,000 mm in the northeast (WB, 2000). About 80 percent of the total rainfall occurs during the monsoon from June to September. In the post-monsoon (October -November) and winter period (December - February) only 10 percent of the annual rainfall is available (WB, 2000). Rainfall is extremely unreliable in the subsequent pre-monsoon period (March - May). On an average there is about 10 percent of the annual rainfall in this period (WB, 2000). On the whole there is a seasonal lack of water depending on the presence and the duration of the monsoon. Water is very scarce in the south and northwest region of Bangladesh during the winter.

Being a country of 140 million inhabitants, agriculture is still the major water using sector for surface and groundwater irrigation with rice cultivation, the single most important activity in the economy. In winter more than 70 percent of crop production is *boro* rice. *Boro* rice is a major food crop which uses up a lot of water per hectare (ha) in the production process. According to one estimate of Biswas and Mondol (1993) it is 11, 500 m³ per ha. Demand for both surface and groundwater for irrigation is on the rise in the dry season which is 58.6 percent of the total demand for water (Chowdhury, 2008) in order to feed a growing population where at least 40 million people do not have a square meal.

Between 1944 and 1999 BWDB spent more than US$1700 m on flood control drainage irrigation (FCDI) projects (WB, 2006). Recently BWDB has handed over these projects to water user groups (mainly medium and large farmers) on the basis of average pricing. This gives rise to a conflict of interests between very small, small farmers and the WUG and hence it is not functional as it should be. More over water has many other uses in the society, fisheries, navigation, mangrove forests, river morphology and not to mention household and industrial uses. Therefore during the dry season water has a high opportunity cost due to competition from all the uses in addition to upstream intervention. It is imperative that farmers pay the true opportunity costs of irrigation water from the perspective of sustainable water use.

International Rice Research Institute (IRRI) conducted an Agricultural Household Survey in Bangladesh in 2000, 2004 and 2008 for 3 crop seasons and collected data on costs of inputs (including irrigation water) and returns on investment from a nationally representative sample of 1880 farm households from 62 villages belonging to 57 of 64 districts of Bangladesh. But the data do not have any information on the volume of irrigation water used in the fields or price per unit. Irrigation is the total irrigation costs measured in BDT (Bangladesh Taka) and we do not know the price per unit of water or the number of hours the pump is used for pumping water. The data available are for total cost of irrigation per household. The costs of irrigation are basically the costs of pumping water. Farmers mainly use low lift pumps for pumping water from surface water sources and shallow and deep tube wells from aquifers and groundwater. I conducted a focussed group discussion in the Northwest region to validate the IRRI data findings about irrigation costs and gather some information about the per unit water costs/prices they bear/pay for using different types of irrigation. Focussed group discussion is useful in case of small samples as opposed to other methods for gathering information within a short time.

## 2. Irrigation institutions

The 5 types of irrigation systems are traditional or local method, canal irrigation project of the government, low lift pump, shallow and deep tube well. When surface water was abundant farmers solely depended on rivers, canals and ponds to irrigate their fields with traditional local methods where the maintenance cost of the apparatus and labour charges when required constituted the costs of irrigation. With the growing population and the introduction of high yielding varieties of rice the government built huge surface water irrigation projects to take care of dry season irrigation. The cost of irrigation became the maintenance of the field channels from the tertiary outlets to the farmers' fields. With the scarcity of surface water in the rivers and canals and advent of groundwater irrigation farmers started paying for pumping water which consists of maintenance of the pumps, fuel cost (electricity or diesel) and the salary of the pump mechanic if required. Farmers use low lift pump to pump water from surface water sources and shallow or deep tube wells for groundwater which are known as minor irrigation devices. The use of low lift pump is limited by the availability of surface water in the canals and rivers during the dry season. Most deep tube wells are government owned and maintained by the public authority. The rest are run on a cooperative or joint ownership. Since investment in deep tube well is lumpy in nature farmers prefer shallow tube wells. Therefore the public sector irrigation institutions are the ones that are taking care of canal irrigation projects of the government and will be discussed in section 2.1. The private sector is the groundwater irrigation using minor irrigation devices to be dealt with in section 2.2.

## 2.1 Public sector

In Bangladesh irrigation is mainly for *boro* rice production in addition to wheat and some other winter crops. Since the 1950s more than 600 water resources schemes have been completed (WB, 2006). These projects ranged from single structure schemes with an impacted area of less than 1,000 ha to large scale multipurpose schemes potentially impacting as much as 100,000 ha. Most were designed to provide flood control, drainage, irrigation or some combination of these. In the late 1950s the government emphasised large scale surface water development projects. Some of the biggest surface irrigation projects are GK (Ganges-Kabodak) project, Pabna Irrigation project, Meghna-Dhonogoda project, Chandpur Irrigation Project, Karnaphuli project, Kaptai project, DND project, Narayanganj-Narshingdi Project, Teesta Project etc. In the mid 80s there was a major breakthrough in HYV (high yielding varieties) rice cultivation through the innovation of Bangladesh Agricultural Research Institute (BARI) and Bangladesh Rice Research Institute (BRRI) and which was possible due to large surface water irrigation projects.

These systems are being run by public agencies like BWDB, Bangladesh Rural Development Board (BRDB) etc. at the district and Upazila level. BWDB is the major public sector agency under the Ministry of Water Resources responsible for planning and execution of over 400 projects developing flood control, drainage and surface water irrigation projects. It also shares an interest in groundwater irrigation as well as in minor surface irrigation with BRDB, Bangladesh Agriculture Development Corporation (BADC) and Local Government Engineering Department (LGED). Earlier, this organization was pioneer in tapping ground water for irrigation in the northern Bangladesh; but subsequently its role in groundwater development was overtaken by BADC and later on by private sector. Major investments in the water sector are made by the Ministry of Water through BWDB and by the Ministry of Local Government and Rural Development through its LGED.

### Participatory water management

In 1994 the Ministry of Water Resources formulated guidelines for people's participation in water development projects to involve local people in water resource projects with the help of officials and experts from BWDB, LGED, Water Resources Planning Organisation (WARPO),

BADC, Department of Agricultural Extension (DAE), Department of Environment (DOE), Department of Forestry (DOF) and the Department of Livestock (DOL).

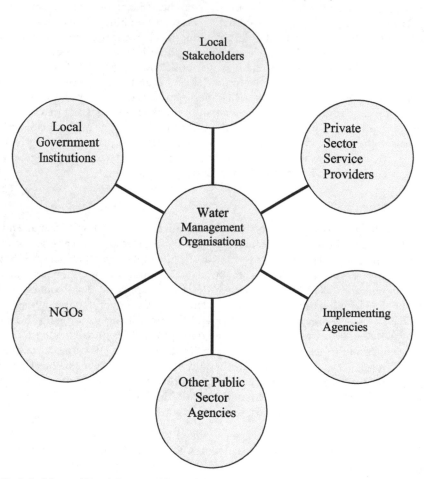

Fig. 1. Stakeholders of Participatory Water Management

Source: Ministry of Water Resources, GOB (2001)

Implementing Agencies are BADC, BWDB, LGED, Barind Multipurpose Development Authority (BMDA) and DOF. Other public sector agencies include DAE, BRDB, Forest Department (FD), DOE, BIWTA, Bangladesh Inland Water Transport Corporation (BIWTC), Department of Cooperatives (DOC), Department of Livestock Services (DLS), Ministry of Land (MOL) and Ministry of Women and Children Affairs (MWCA).

The principle that community resources must be managed by the community concerned along with local government institutions guides participatory water management. The National Water Policy emphasises the issue of participatory water management through planning, stakeholder participation, public and private management, economic and financial management and institutional policy. Local Governments (Parishads) are the principal agencies for coordinating the design, planning, implementation, and operation and maintenance (O&M) of publicly funded surface water resources development projects. The participatory process also depends on NGOs and community level self-help groups (private).

The institutional framework of participatory irrigation management (PIM) in which the local stakeholders participate commenced in 1995, introduced a three-tier management structure for irrigation systems. This involved creating tertiary-level Water Management Groups (WMGs; each consisting of nine members—three from each of three farm-size categories: 'large', 'medium' and 'small'); secondary level Water Management Associations (WMAs; consisting of 10-15 WMGs); and a Water Management Federation (WMF) at the highest level of a system. Water Management Organisations are responsible for planning, implementing, operating as well as maintaining local water schemes in a sustainable way. They also contribute towards the capital and operating costs of the scheme as decided by the Government or on a voluntary basis acting in their own interest.

Ownership of flood control drainage irrigation (FCDI) projects with command area of 1000 ha or less is gradually transferred to the local governments with the ones that are satisfactorily managed and operated by the beneficiary/community organisations. The management of public water schemes with command area of up to 5000 ha are gradually made over to local and community organisations and their O&M are to be financed by local resources. Public water schemes with command area over 5000 ha are gradually given to private management through leasing, concession, or management contract under open competitive bidding or jointly managed by the project implementing agency along with local government and community organisations.

Appropriate public and private organisations provide information and training to the local community organisations for efficient management of water resources. For minor irrigation stakeholders participation is confirmed by their willingness to commit to financial contributions before receiving services. In case of FCDI projects water rates are charged for O&M as per government rules. Water charges realised from beneficiaries for O&M in a project are retained locally for the provision of services within that project. Some use an informal structure while others use a formal structure with legally registered organisations according to the guidelines for participatory water management.

BADC during its programme to expand groundwater irrigation required WUGs to be formed. The Barind Multipurpose Area Development Project has successfully used cooperatives for water management. The Ministry of Agriculture has a shared experience in participatory management under National Minor Irrigation Project (NMIP) where beneficiaries voluntarily re-excavated canals to support LLP irrigation. The Department of Public Health Engineering has also introduced participatory management to support rural water supply and sanitation program. An important development in participatory management has been in Small Scale Water Resources Development Project of LGED. The beneficiaries have participated in the water management projects right through its initiation by making a percentage payment toward investment and for operating and managing the project entirely by them.

Much effort was given in the past two decades to review stakeholders' participation under the Dutch aided Early Implementation FCD projects, and IDA (International Development

Assistance)/CIDA (Canadian International Development Assistance) assisted small-scale FCDI projects. This initiative was later enforced through Flood Action Plan (FAP) studies and the System Rehabilitation Project (SRP) of the BWDB. Lately, the LGED started developing small irrigation projects having an area less than 1000 ha in order to improve efficiency and coordinate better with other infrastructure building efforts.

However, except for a very few, the experience of participatory management has been a mixed one. Progress in creating and developing these user groups has been slow, and the irrigation sector continues to be managed at all levels by public-sector agencies. Perhaps, it is due to some gap at the initial stage of the project preparation where it might not have been possible to involve the beneficiaries at all stages of project cycle. Although the guidelines for participatory water management are an excellent starting point to promote local stakeholder involvement in water management infrastructure these are inadequate in promoting meaningful participation (WB, 2006). Rather than establishing the mechanisms to improve agencies ability to respond to local stakeholders, the guidelines encourage the participation from the perspective of achieving the objectives of the executing agencies.

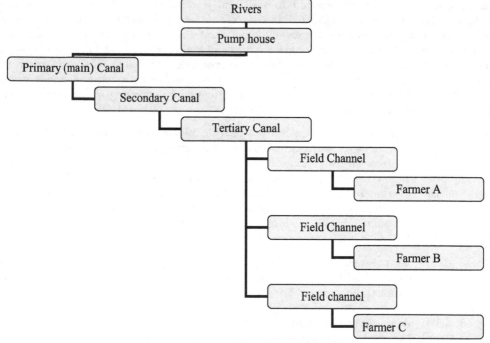

Fig. 2. National Surface Water Irrigation Distribution Systems

Individual farmers get water in their fields through the field channels. Farmers are responsible for maintaining the field channels from the tertiary canals.

**Some case studies**

There are 3 types of institutional models for large scale surface water irrigation schemes in Bangladesh. The BWDB controls Meghna-Dhonagoda Project and Pabna Irrigation Project down to the tertiary outlets and work with the registered user groups organised in several tiers under the cooperative system. The user groups collect water charges and participate in overall scheme management and operation and maintenance. Asian Development Bank (ADB) supports these schemes. Also irrigation schemes are managed by an authority like BMDA.

**Pabna Irrigation and Rural Development Project** (PIRDP) is located on the floodplain of the Brahmaputra and Hurasagar rivers (west-central Bangladesh). This project aimed to provide flood control, drainage, and irrigation facilities. The total project area is 186,000 ha and the area irrigated is 145,000 ha. 77 percent of the total rainfall occurs between mid June and mid October. Rice is the major crop grown on 64 percent of the total cropped area. Other crops are pulses, potato, jute, sugarcane, vegetables, onion, wheat and oilseeds. Land is inequitably distributed average landholding is 0.92 ha. 6% of landowning households own 25% of the available land and 14% households are landless. 78% households own 1 ha of land or less. The average annual incomes of small-, medium-and large-scale farmers are USD 487, USD 846 and USD 1,347, respectively. The net value of crops produced per ha is USD 203. The net benefit of irrigation is USD 125/ha. The irrigation benefits mainly result from access to water and use of HYV technology on the irrigated land. Farmers with very small landholdings (less than 1 ha) obtained higher yields per ha than larger landholders. This is because the land-poor farmers use the available water more efficiently, grow HYVs of rice on a greater proportion of their land, and irrigate more intensively than farmers with more land.

Only 2 of the 365 WMGs formed have been registered with the government. These groups do not assess or collect irrigation charges, or dictate how the revenues are spent. These functions are still being performed by the BWDB. Since user groups have not yet taken on this role, as originally envisioned, collection rates remain very low, 9% of the target. Water charges are not based on the volume of water used, but on (1) the area irrigated (irrespective of the size of the farm), and (2) the type of crop grown (depending upon its water requirements). Charges average around BDT 540/acre (USD 22.39/ha).

The **Thakurgaon Irrigation Project** consists of a cluster of deep tube wells and is transferred to a private company established by the local stakeholders who sells water on a pay as you use system. The project is now running successfully where payments are made in advance to pump operators at a specified unit rate and they provide water to the farmers within the command area according to the terms of payment on a first come first served basis. The system provides for long term management, operation and maintenance for the equipment.

In case of **Muhuri Irrigation Project** there is no functioning organisation of local stakeholders and also there is no opportunity for minor irrigation development. Therefore large scale surface water management infrastructure is the only alternative and the project infrastructure provides farmers with flood control and access to surface water for irrigation through a network of secondary channels. As the system does not extend to the tertiary level and beyond development costs are lower but farmers are responsible for obtaining, operating and maintaining their own pumps to abstract water from these channels and for the associated field distribution systems. However the major infrastructure constructed was of good quality and continues to function effectively and provide the anticipated benefits.

At present the best institutional model is the ADB financed Small Scale Water Resources Development Sector Project which provides flood control, drainage or irrigation infrastructure to subproject areas less than 1000 ha. These projects rely heavily on local stakeholders' initiative to identify interventions, ratify engineering designs, demonstrate commitment to operate and maintain the infrastructure by contributing a specified amount of funds in advance of physical construction. The projects are implemented by the LGED with support from the ADB and the Government of Netherlands. In some cases low quality infrastructure, shortcuts in the development process and inconsistency are responsible for failure. In order to make water management investments pro local stakeholders' financial responsibility, decision-making authority and accountability must be transferred to local stakeholders and their representatives – the local government. Investments have to be structured so that service agencies are obligated to cater the end users requirements.

## 2.2 Private sector groundwater irrigation

After 1974 many surface water irrigation projects like the GK project became ineffective due to operation of the Farakka Barrage. Also these projects had long gestation periods, suffered from management and maintenance problems and were unpopular with farmers because the distribution canals took up scarce land. In 1980s, there was a surge in private sector involvement in ground water extraction mostly by shallow tube wells. At present the role of BADC is very minimal. Over time the government shifted emphasis to small scale projects, fielding power pumps to lift surface water from creeks and canals and tube wells for extraction of groundwater. Since then many farmers switched to 2 rice crops and vegetables and other crops which require much less water instead of 3 rice crops during the *boro* season and also moved to shallow tube wells instead of deep tube wells. This was the time when government emphasised groundwater irrigation.

Private sector groundwater development for irrigation has been instrumental in expanding agricultural output. It is also recognised that the development of groundwater for irrigation has virtually required no public sector financing in contrast to a huge capital cost of surface water systems ranging from USD 500 to 1800 per ha. Capital investment in the Teesta Project was about USD 250 m in 1985 to develop 100,000 ha of irrigation. It is further recognised that farmers readily mobilise the financial and technical resources to operate and maintain groundwater irrigation infrastructure whereas in surface water systems, in most cases the cost of irrigation user fee collection has exceeded the fees collected. The NWMP also notes that irrigation intensities are low on the 15 major existing irrigation schemes. On the whole, the large scale surface water irrigation projects do not have a good performance record. Minor irrigation is a source of net revenue (diesel tax revenue less electricity subsidies). Diesel is used to power about 90 percent of pump sets and irrigates 70 percent of the area. Electricity powers the rest. Present price policy subsidise electricity while diesel is taxed. As the end user farmers own the equipment and usually provide irrigation water to neighbouring plots at competitive prices. Infrastructure quality, management and operation, maintenance and benefits are not issues. The individual owners ensure that their requirements for the equipment and its use are met.

Irrigation has expanded to more than 50 percent of cultivated land (Hossain et al, 2007) and is provided through minor irrigation devices such as low lift pumps (LLP), shallow tube wells (STW) and deep tube wells (DTW). Initially LLPs, DTWs and STWs were supplied by Bangladesh Agricultural Development Corporation (BADC) a public sector organisation.

Since the early 1980s the government has privatised the procurement and distribution of minor irrigation equipments, reduced import duties and removed the restriction on the standardisation of irrigation equipments. As a result farmers have made substantial investment in shallow tube wells and power pumps contributing to rapid expansion of irrigation facilities since the mid-1980s. The area irrigated by tube wells expanded from 53, 000 ha in 1973 to 3.3 m ha in 2000 (Hossain et al, 2007). Shallow tube wells and power pumps accounted for 71 percent of total irrigated area in 2000. Small scale private investment on low lift pumps and tube wells and development of a competitive market for water transactions from tube wells to small and marginal farmers accelerated rapid expansion of irrigation. The diffusion of modern variety *boro* rice is strongly related to groundwater irrigation expansion. After the privatisation of minor irrigation LLPs and STWs became more popular compared to DTWs which required high initial investments.

Bangladesh is a delta composed of ridges and troughs. Soil texture is heavy on troughs and light on ridges. Plots on high land have different cropping patterns due to different soils, flooding regimes and access and returns to irrigation sources compared to medium and lowland (Palmer-Jones, 2001). Water loss from canals is high on more elevated land with light textured soils. Low land is suitable for rice cultivation since it is permanently waterlogged. Returns to irrigation are spatially variable due to soil and hydro-geological characteristics.

Modern irrigation mainly consists of STW which has almost replaced LLPs and is gradually displacing DTWs which are economically and socially unfavourable. The remaining DTWs although initially set up as either formal or informal cooperatives have become privately owned.

There are various institutional forms of ownership and management of STW. Many STWs are jointly owned by relatives, neighbours or friends. Usually a pump operator is engaged for the whole irrigation season who may also be the owner or one of the users for a fixed seasonal fee in cash or kind. In many places water is paid by one fourth of the gross crop harvested and delivered to the tube well owner. A large part of capital costs and operation and maintenance costs come from outside the village like business, service and remittances.

In Bangladesh informal water markets for irrigation have developed quickly with the rapid expansion of tube well irrigation over the last decade. In case of shallow and deep tube wells, the owners of the irrigation equipment enter into deals for irrigation services with neighboring farmers in addition to using the equipment for irrigating their own land. With the expansion of water markets in the private sector, the pricing system has also undergone changes to suit varying circumstances. There is no single rate or uniform method for payment of irrigation water. Per hectare water rates vary not only from one area to another but also depend on the type of well within a particular area (Biswas and Mandal, 1993).

In the initial stage, the most common practice was sharing one-fourth of the harvest with the owner of the equipment in exchange for water. That gave way to a flat seasonal fee, the rate depending on the availability of electricity and the price of diesel. In recent years, the market has moved toward fees per hour of tube well operation. In Bangladesh, the major source of irrigation is the shallow tube wells and power pumps mostly run by diesel as many places in rural Bangladesh still do not have electricity connection. Diesel pumps usually have higher costs and lower water extraction capacity than electricity operated pumps (Wadud and White, 2002). Diesel being a major agricultural input in the cultivation of *boro* rice, the cost of *boro* cultivation is very sensitive to the price of diesel.

| Type | STW | | DTW | |
|---|---|---|---|---|
| Name | Shallow tube well | | Deep tube well | |
| Description | Shallow well with suction mode pump | | Usually turbine type pump (in large diameter) well 150-300m deep | |
| Energy | Diesel | Electricity | Diesel | Electricity |
| Nominal Capacity (litres/second) | 12 | 12 | 50 | 50 |
| Overall efficiency | 25% | 35% | 35% | 35% |
| Energy Cost (BDT) per ha | 4,040 | 1,570 | 5,410 | 2,950 |
| Total cost (BDT) per ha | 6,990 | 3,770 | 12,940 | 8,930 |

Note: Irrigated area assumes 10 hours pumping daily and energy costs are based on diesel fuel costs of BDT 14/litre and electricity at BDT 25/KWh. Capital cost is annual equivalent capital cost at 12% discount rate divided by the command area. 1 USD = 69 BDT (Bangladesh Taka).
Source: WARPO (1999).

Table 1. Estimated Total Costs for Different Well Technologies

For Bangladesh the cost of production is higher for the *boro* rice than for the *aman* variety of rice. A major factor behind the high unit cost of *boro* rice cultivation in Bangladesh is the high cost of irrigation compared to the other countries in the region. Bangladeshi farmers have to spend about USD 51 in irrigating one-hectare land whereas the irrigation costs are about USD 32 in Punjab, India (Hossain and Deb, 2003). The cost of MV *boro* irrigation is even higher in Bangladesh; it is USD 117.6 per ha (Hossain and Deb, 2003). In Bangladesh, irrigation costs account for 28 percent of the variable costs of rice cultivation.

Further, in Bangladesh there has been a rising dependence on groundwater due to lack of surface water in the recent past. Overexploitation of groundwater for irrigation and other purposes has lowered the water table in many parts of the country below the suction level of the tube wells. The result is the increased costs for irrigation. Based upon the field study, NWMP (National Water Management Plan) estimates of operating costs for supplying 11,000m$^3$ of water (the typical gross demand for 1 ha of *boro* rice) are given in Table 1. The costs of diesel operation are substantially higher than electricity. Part of this is due to the generally lower efficiency of diesel-powered pump sets, but the major cause is that diesel fuel is taxed whereas electricity is charged at a price lower than its production cost. On the other hand, India provides heavy subsidy on electricity that lowers the cost of irrigation. In Indian Punjab electricity is provided free for tube well irrigation and the farmers are also provided free water from irrigation canals.

The present government policy for water management is the conjunctive use of surface and groundwater. Government and donors agree on policies to promote the expansion of irrigation from groundwater using tube wells provided by the private sector. According to the National Water Policy water should be used most economically. Farmers do not pay for volume used for irrigation but pay for the operation and maintenance. They pay for digging canals, diesel, electricity, labour charges for running pumps etc. At present 90 percent of total irrigation of 4.5 m ha is from groundwater, the rest is from surface water mainly

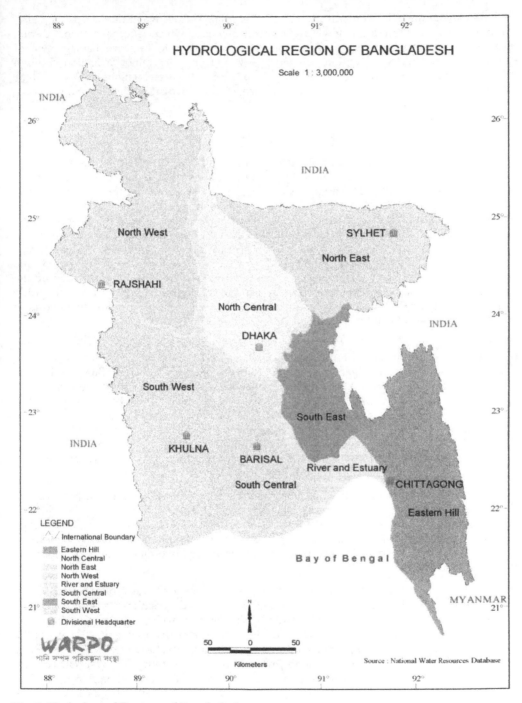

Fig. 3. Hydrological Regions of Bangladesh

because during the irrigation season (winter) there is no surface water due to upstream use. Mining and over-extraction is lowering groundwater table which is recharged only during floods. The National Water Management Plan emphasises to improve management of existing surface water irrigation schemes. It would support minor irrigation through improvement of shallow tube well and deep tube well diesel engine fuel efficiency, introduction of lower cost pump sets and improvement of irrigation distribution to reduce losses in water short areas.

## 3. Study area: Northwest region

The objective of this study is to measure the volume of irrigation water and find the actual price of irrigation water per unit used by farmers for alternative modes of irrigation in the Northwest region (NW) of Bangladesh. Water is most scarce in the Northwest and Southwest region during the dry season due to low annual rainfall. The problem is more complicated in the Southwest region due to water logging and salinity intrusion from the Bay of Bengal in addition to low annual rainfall. Hence I have chosen the Northwest region in order to isolate the impact of dry season scarcity of water from other seasonal and environmental problems. Previous studies (Chowdhury, 2005; Linde-Rahr, 2005) used total irrigation costs instead of water as a physical input since there was no information on the price per unit of water or volume of water used in the dataset. Through focussed group discussion we collect data on volume of water used in irrigation in addition to costs of irrigation in 3 districts of the NW region.

Bangladesh is an agricultural country divided into 7 hydrological regions. Northwest region encompasses the Rajshahi Administrative Division of 16 districts and is bounded by the Brahmaputra and Ganges rivers. Apart from Rajshahi Rangpur, Dinajpur, Bogra and Pabna are the main urban centres. Average rainfall is about 1700 mm but its south western part in the Barind zone is one of the driest in Bangladesh with average rainfall below 1400 mm. The Barind Tract is the driest part of the region where surface water supply is very limited. The tract extends over Rajshahi, Dinajpur, Rangpur and Bogra districts of Bangladesh and Maldah district of West Bengal, India. The temperature varies between 6 and 44 degree Celsius. Apart from the monsoon from mid June to mid October the climate is very dry. The high Barind is the only elevated land.

The Barind Tract is distinguished by hard red soils and older alluvial deposits which are different from other parts of Bangladesh. The main clay minerals are kaolinite, chlorite, smectite, and mica-smectite interlayered phases. The Barind clay contains an average total organic carbon content of 0.05%. The agro climatic conditions of the Barind region are highly favourable for irrigation but required an enormous government support before this potential could be realised.

Irrigation in the northwest region is almost entirely dependent on groundwater. Its net cultivable area is 2.35 m ha. The region is highly developed agriculturally with the largest irrigated area of all regions supplied mainly by shallow tube wells. Initial experiments with tube well irrigation began in the 1960s with the Thakurgaon Deep Tube well Project in the northwest. Due to STW pumping irrigation seasonal water table decline is widespread. The southern part of this region is very flood prone. Some of the country's biggest flood control drainage and irrigation schemes are located in this area. These are Teesta Barrage, Pabna Irrigation Project, and the Barind Multipurpose Development Authority (BMDA). BMDA includes a deep tube well irrigation component.

In the 1990s except the special case of the Barind project government withdrew any involvement in groundwater irrigation. An independent autonomous body, the Barind Multipurpose Development Authority was established to implement the projects under the direct supervision of the Ministry of Agriculture. The main objective was to improve the quality of life of the people by ensuring year round irrigation, augmentation of surface water resources, improvement of agricultural support services etc. Currently the Barind aquifer is supplying sufficient water for extensive irrigation. The project is developed with meagre international support either in terms of experts or finance. There are active staffs, good management and farmers participate enthusiastically because their well being is greatly enhanced by these developments.

| Variety | Classification | Growth duration (days) | Yield per ha (ton) |
|---|---|---|---|
| BR29 | medium fine | 165 | 7.5 |
| BR16/Shahi balam | fine | 165 | 6 |
| BR14/Gazi | medium coarse | 160 | 6 |
| BR11/Mukta | medium coarse | 145 | 6.5 |
| BR28 | medium fine | 140 | 5 |
| BR1/Chandina/Chaina | coarse | 150 | 5.5 |
| BR9 | medium coarse | 155 | 6 |
| BR36 | fine | 140 | 5 |

Table 2. *Boro* Rice Varieties

There are 18 varieties of *boro* rice being cultivated in the Northwest region. I classify them into 3 broad categories fine, medium and coarse with the help of Bangladesh Rice Research Institute Scientists. Although evapo-transpiration is the true water requirement for crop growth crop water requirement for rice equals seepage, percolation and evapo-transpiration. Rice plants require continuous water in addition to land preparation. About 200 mm (heavy soil) to 250 mm (light soil) water is required for land preparation in *boro* rice cultivation. Therefore irrigation water requirement for *boro* rice is 3 times higher than non-rice crop like wheat and maize. Usually the seepage and percolation rate in rice fields varies from 4 to 8 mm per day (Rashid, 2008). Seepage and percolation rates are higher for light than heavy soil.

The rate of water requirement varies with atmospheric condition (temperature, rainfall), soil type, crop age, duration of the crop growth, land elevation and water management status in the plot. No irrigation is required before 10-15 days of harvest. Water balance studies show that much more water is supplied by irrigation than is required for evapotranspiration, seepage and percolation. According to one BRRI study 4,000 litres of water is used as irrigation for per kg *boro* rice in farmers' field compared to 2,000 litres in an experimental plot. Irrigation water requirement for *boro* rice production also varies with the varieties of longer duration.

| Rice variety | Growth duration (days) | Water required for heavy soil (mm) | Water required for light soil (mm) |
|---|---|---|---|
| BR28 (medium duration) | 140 | 995 | 1355 |
| BR29 (long duration) | 160 | 1205 | 1640 |

Table 3. Water Requirement for *Boro* Rice under Continuous Standing Water

| | *Boro* rice | Wheat/non-rice cultivation | STW (shallow tube well) | DTW (deep tube well) |
|---|---|---|---|---|
| Northwest region | BDT 7000-8000 per ha | BDT 50-60 per hour | | |
| Dinajpur | BDT 2500 per ha* | BDT 35-45 per hour * | BDT 30-60 per hour for wheat | |
| Rajshahi | BDT 7000 per ha** | BDT 1852 per ha** | BDT 50 per hour *** | BDT 75-80 per hour*** |

Source: As available from various BARC (Bangladesh Agricultural Research Council) Reports, 2001.
* in a village in Dinajpur district (NW region).
** Barind area (NW) in 1998-99.
*** In *boro* season.

Table 4. Cost of Irrigation in BDT for different Districts of NW region

Irrigation cost is different for different crops. Most popular is one-fourth of the crop share for irrigation in paddy cultivation. Pump owners provide the fuel and oil. Since it leads to a flat seasonal fee and sometimes overuse of water resulting in higher marginal cost per unit water extraction, it is important that we have information on volume of water used in irrigation in farmers' field.

## 4. Methodology: Focussed group discussion

International Rice Research Institute (IRRI) conducted an Agricultural Household Survey in Bangladesh in 2000, 2004 and 2008 for 3 crop seasons and collected data on costs of inputs (including irrigation water) and returns on investment from a nationally representative sample of 1880 farm households from 62 villages belonging to 57 of 64 districts of Bangladesh. But the data do not have any information on the volume of irrigation water used in the fields or price per unit. Irrigation is the total irrigation costs measured in BDT (Bangladesh Taka) and we do not know the price per unit of water or the number of hours the pump is used for pumping water. The data available are for total cost of irrigation per household. The costs of irrigation are basically the costs of pumping water. Farmers mainly use low lift pumps for pumping water from surface water sources and shallow and deep tube wells from aquifers and groundwater. I conducted a focussed group discussion in the Northwest region to validate the IRRI data findings about irrigation costs and gather some information about the per unit water costs/prices they bear/pay for using different types of irrigation mentioned earlier. Focussed group discussion is useful in case of small samples as opposed to other methods for gathering information within a short time.

The task was to collect some samples of data on water costs/prices per acre for *boro* rice from representative farmers who are using 5 different types of irrigation options in the NW region. These irrigation modes are low lift pump, shallow tube well, deep tube well, canal irrigation project of the government and local or traditional irrigation system. Data were collected on farmers' use of irrigation water, volume, price they pay for hiring pumps, if they own pumps its cost of operation run by diesel or electricity, daily hours of pumping water during the irrigation season, capacity of pumps to extract water per second in litres etc. If it is a government owned irrigation project then the same information is collected on low lift pumps and other modes of irrigation the farmers are using. I also gathered information on the water table there. Three districts are Rangpur, Rajshahi and Pabna.

Rangpur represents northern range or highland, Rajshahi is from Barind Tract representing medium high land and Pabna represents lowland. In Rangpur besides the private sector there is Teesta irrigation project. In Pabna district there are government run irrigation projects like Pabna irrigation project plus privately owned pumps. Pabna Irrigation Project is mainly based on surface water irrigation. In Rajshahi the major irrigation projects are run by the Barind Multipurpose Development Authority. From these 3 districts 15 farmers using 5 different types of irrigation options were interviewed. Thus the focussed group discussion is for 15 farmers. The results are summarised in the following table.

| District | Village | Respondent | Land under boro rice | Harvest | Irrigation type | No of irrigation | Inches of water per bigha | Cost of Irrigation | Energy source |
|---|---|---|---|---|---|---|---|---|---|
| Rajshahi | Achua taltola | 1 | 1 acre | 45 maunds | Deep tube well | 13 | 2.5 | BDT 3600 per acre | Electricity |
| Rajshahi | Moishalbari | 2 | 1.67 acre | 75 maunds | Shallow tube well | 15 | 1.5 | BDT 3600 per acre | Electricity |
| Rajshahi | Moishalbari | 3 | 2 acre | 72 maunds | Low lift pump | 15 | 1.5 | BDT 4500 per acre | Electricity |
| Pabna | Bhabanipur | 4 | 0.33 acre | 25 maunds | Deep tube well | 65 | 3 | 1/4th | Electricity |
| Pabna | Jobedpur | 5 | 3.33 acre | 200 maunds | Deep tube well | 120 | 3 | BDT 6000 per acre | Electricity |
| Pabna | Jobedpur | 6 | 1 acre | 36 maunds | Shallow tube well | 75 | 2 | 1/4th | Electricity |
| Pabna | Sonatola | 7 | 3.67 acre | 308 maunds | Low lift pump | 105 | 3 | BDT 3600 per acre per m+1500 | Diesel |
| Pabna | Sonatola | 8 | 2.33 acre | 180 maunds | Traditional method | 100 | 2.5 | BDT 1500 per month | |
| Pabna | Nandanpur | 9 | 1.33 acre | 65 maunds | Canal irrigation | 65 | 4 | BDT 540 per acre | |
| Rangpur | Shyampur | 10 | 5 acre | 60 maunds | Deep tube well | 20 | 2 | BDT 700 per acre+m salary | Electricity |
| Rangpur | Pakuriasharif | 11 | 1 acre | 100 maunds | Deep tube well | 30 | 3 | BDT 3181.81 per acre | Electricity |
| Rangpur | Godadhar | 12 | 5.5 acre | 280 maunds | Shallow tube well | 35 | 3 | BDT 2800 per acre | Electricity |
| Rangpur | Nabanidas | 13 | 3 acre | 225 maunds | Shallow tube well | 45 | 4 | BDT 4545.45 per acre* | Electricity and diesel |
| Rangpur | Dighaltari | 14 | 2 acre | 100 maunds | Shallow tube well | 64 | 5 | BDT 3600 per acre per week | Diesel |
| Rangpur | Dighaltari | 15 | 3.33 acre | 250 maunds | Canal irrigation | 30 | 5 | BDT 900 per acre | Hand tube well |

*BDT 2727.27 per acre if run by electricity
1 maund =28 kg
1 hectare = 2.47 acre
1 acre = 3 bigha
1 bigha-inch = 20,588 litres

Table 5. Results from my Field Study in the Northwest Region

In Rajshahi most of the irrigation pumps are installed and maintained by the BMDA. We interviewed farmers in 2 villages of Godagari Upazila using STW, DTW and LLP in Sharmongla canal irrigation project. The Barind Authority owns the DTW and the LLP. The LLP user here incurs more cost than the DTW and the STW users per acre. All the pumps are run by electricity. We did not however find anyone using traditional or local irrigation method in Rajshahi.

In Pabna I interviewed 6 farmers and found that two farmers pay the irrigation cost in terms of crop (1/4th). Here we find a very high frequency of irrigation in case of one DTW user and in general all 6 farmers compared to the farmers of Rajshahi and Rangpur. These farmers are producing BR29 and IRRI29 *boro* varieties which have longer duration. The soil quality may be also lighter than that in Rangpur and Rajshahi which requiring more water. DTW has the highest cost of irrigation. Electricity or diesel cost and the monthly wages of the pump operator constitute the total costs of irrigation for those who are using DTW, STW and LLP. Traditional method and canal irrigation system are cheaper modes of irrigation. In Rangpur the frequency of irrigation is found lower than Pabna but higher than in Rajshahi. I interviewed 6 farmers in Rangpur. Here we found an interesting case where one farmer who is running a STW for irrigation is using both electricity and diesel in the absence of electricity. In this case it is noteworthy that irrigation cost is less than double when he has to use diesel as fuel instead of electricity.

### Lessons from the focussed group discussion

According to this field research one can get the information on water volume in two ways. One is from the horsepower of the pump, the number of hours the machine is run and another way is the number of inches of water on the plot. As it is not possible to know the level of efficiency of the pumps that are operating it is more reliable to measure the quantity of water used from the number of inches of standing irrigation water on the field. We did not however find anyone using traditional or local irrigation method in Rajshahi.

It is obvious from these interviews that farmers cannot reveal the amount of water withdrawn per minute or hour when they are running the pumps to irrigate their fields. However one can estimate the amount of water from the inches of standing water on their fields each time they irrigate their plots which is clear from this small sample survey.

Farmers prefer traditional method to government canal irrigation project to STW to LLP to DTW the cost of whichever is less. Electricity run pumps always cost a bit more than half the amount of diesel run ones. The use of traditional/local method, canal irrigation project and LLP depends on the availability and proximity of surface water. When groundwater is the only choice STW is preferred to DTW from the view point of least cost as investment in DTW is lumpy in nature. DTW is beyond the means of the poor mass of the farmers to be single owners. But in some cases DTW can be cheaper to an individual due to large economies of scale. Government run/maintained DTWs cost less to farmers than DTWs run on the basis of joint ownership and the rented ones. Cost of irrigation for pumps is the energy cost and the salary of the pump manager if hired or maintained by the government. The monetary cost has to be weighed against pump management and electricity availability.

For traditional (local) irrigation method the cost of irrigation is the maintenance of the apparatus and the cost of hired labour if any. For the government run canal irrigation methods the farmers usually pay a fixed amount per unit of irrigated land per crop during the season and in case of participatory water management water user groups pay for maintenance of field channels plus management cost in some cases. It varies from case to case. It is more reliable to measure the amount of irrigation water in terms of bigha inch water and the number of times the rice field is irrigated. In many cases farmers or even some pump managers cannot tell the level of efficiency of the pumps they are running. Only farmers who also happen to be pump owners/managers at the same time could give us accurate information about the age and the durability of the irrigation pumps.

## 5. Conclusions

Farmers in Bangladesh do not pay for use of per unit of irrigation water. When surface water was abundant farmers solely depended on rivers, canals and ponds to irrigate their fields with traditional local methods where the maintenance cost of the apparatus and labour charges were the costs of irrigation. For the government run canal irrigation methods the farmers usually pay a fixed amount per unit of irrigated land per crop during the season and in case of participatory water management water user groups pay for maintenance of field channels plus management cost in some cases. It varies from case to case. With the advent of groundwater irrigation cost of irrigation consists of maintenance of the pumps, fuel cost (electricity or diesel) and the salary of the pump mechanic when required. Farmers use low lift pump to pump water from surface water sources and shallow or deep tube wells for groundwater. The use of low lift pump is limited by the availability of surface water in the canals and rivers during the dry season. Since investment in deep tube well is lumpy in nature farmers prefer shallow tube wells. In case of shallow tube wells the energy cost (electricity or diesel) is the main component of irrigation cost. As electricity is not available in all the villages farmers have to depend on diesel to run irrigation pumps to a large extent. Hence the price of diesel in the international market plays a crucial role in cost of irrigation for the private sector.

The main challenge of public sector irrigation institutions is to design proper incentives for all stakeholders to participate in the participatory water management network. Successful water management practice for irrigation will depend on equitable participation of all groups of farmers as water user groups in management and cost recovery. Introduction of rice varieties that require less water for irrigation per ha is mandatory. Government should give incentives or price support for wheat and maize production so that farmers diversify towards these crops that require much less water per ha for irrigation compared to boro rice. In order to run the pumps with electricity stability in power supply is a must that will reduce the cost of irrigation as well as cost of cultivation drastically. In this endeavour there is no alternative to 100 percent rural electrification. More case studies or field research with large samples will demonstrate the actual status of irrigation institutions from the perspective of policy design and implementation specially if successful examples are identified and replicated elsewhere.

## 6. Acknowledgements

This paper has benefited from many comments and suggestions of Professor A K Enamul Haque. I am also grateful to Professors Mustafa Alam, Syed M Ahsan, Dr Priya Shyamsundar, and other SANDEE (South Asian Network on Development and Environmental Economics) Advisors for their comments. Dr Mahabub Hossain, Executive Director, BRAC (Bangladesh Rural Advancement Committee) kindly gave me the permission to rent rooms in their training and resource centres in Rangpur, Rajshahi, and Pabna during my field research in the villages. The study grant from SANDEE is gratefully acknowledged. Finally I express my sincere gratitude to Economic Research Group (ERG) for hosting this study grant.

## 7. Appendix

**Questionnaire**

First part of the questionnaire constitutes socio-economic information of the farmers.
Name of the respondent
Relationship with the household head
Gender
Age
Village
Union
Upazila
District
Level of education
Other occupation of the household and the members
Total land owned by the household
Per capita annual income of the household
Village with electricity connection: yes/no
Second part includes farming and irrigation related information.
Total land cultivated
Terms of lease: own, shared, leased in, leased out
Crops produced
Area under boro rice cultivation
Type of boro rice (variety)
Quantity harvested in kg
Quantity sold in kg
Price obtained per kg
Source of irrigation (own tube well, shared tube well, hired tube well, low lift pump, shallow tube well, deep tube well, government irrigation canal project, local irrigation system)
Type of tube well (submersible or non submersible)
Name of owners (both own and joint)
Year of installation
Depth of bore hole, filter and pump
Depth of water level
Horsepower of the pump
Cost of installation
Whether run by diesel or electricity
Costs of diesel or electricity
How the energy costs are shared
Whether labour/mechanic is required to pump water (yes or no)
If yes, his charges
The depth of water in inches on the ith irrigation

Number of hours on the ith irrigation taken to flood the field to reported number of inches
Terms of irrigation when the source is a shared tube well
Terms of rent when the pump is hired
Distance of the plot from the irrigation source

## 8. References

Ahmad, Q. K., A. K. Biswas, R. Rangachari and M. M. Sainju edited (2001): *Ganges-Brahmaputra-Meghna Region: A Framework for Sustainable Development*, University Press Limited, Dhaka, Bangladesh.

Asian Development Bank (2004): Pro- poor Interventions in Irrigated Agriculture Issues, Options and Proposed Actions, Bangladesh, Document of the Asian Development Bank.

Banerji, A. and J.V. Meenakshi (2006): Groundwater Irrigation in India: Institutions and Markets, SANDEE Working Paper 19:06.

BARC (2001a): Bangladesh Agricultural Research Council, Contract Research Project on Improvement of the Water Resources Management for a Sustainable Rice-Wheat system in Dinajpur and Jessore Areas, Irrigation and Water Management Division, Bangladesh Agricultural Research Institute, Gazipur-1701, November 2001.

BARC (2001b): Bangladesh Agricultural Research Council, Contract Research Project on Improvement of the Water Resources Management for a Sustainable Rice-Wheat system in Barind Area, Irrigation and Water Management Division, Bangladesh Rice Research Institute, Gazipur-1701, December 2001.

Biswas, M. R. and M. A. S. Mandal (1993): *Irrigation Management for Crop Diversification in Bangladesh*, University Press Limited, Dhaka, Bangladesh.

Chowdhury, N. T. (2005): "The Economic Value of Water in the Ganges-Brahmaputra-Meghna River Basin", Beijer Discussion Paper 202, October 2005, Beijer International Institute of Ecological Economics, the Royal Swedish Academy of Sciences, Stockholm, Sweden. Available at www.beijer.kva.se/publications/pdf-archive/Disc202.pdf

Chowdhury, N. T. (2010): "Water Management in Bangladesh: An Analytical Review", *Water Policy, 12(1)*, 32-51.

Government of Bangladesh (2005): Unlocking the Potential National Strategy for Accelerated Poverty Reduction, General Economics Division, Planning Commission, and Government of the People's Republic of Bangladesh.

Hossain, M., D. Lewis, M. Bose and A. Chowdhury (2007): Rice Research, Technological Progress and Poverty: The Bangladesh Case in *Agricultural Research, Livelihoods, and Poverty, Studies of Economic and Social Impacts in Six Countries*. The Johns Hopkins University Press, Baltimore, USA.

Hossain, M. and U. K. Deb (2003): Trade Liberalization and the Crop Sector in Bangladesh, Paper 23, the Centre for Policy Dialogue, Dhaka, Bangladesh.

Kijne, J.W., T. P. Tuong, J. Bennett, B. Bouman and T. Oweis (2003): Ensuring Food Security via Improvement in Crop Water Productivity, Challenge Program on Water and Food Background Paper 1.

Linde-Rahr, M. (2002): Household Economics of Agriculture and Forestry in Rural Vietnam, Ph. D thesis, Gothenburg University, Gothenburg, Sweden.

Ministry of Water Resources, GOB (2001): Guidelines for Participatory Water Management, Dhaka, Bangladesh.

Molden, D. and R. Sakthivadivel (1999): Water Accounting to Assess Use and Productivity of Water, *Water Resources Development*, *15*, 55-71.

Mukherji, A. (2004): "Groundwater Markets in Ganga-Meghna-Brahmaputra Basin: Theory and Evidence", *Economic and Political Weekly*, 3514-3520.

Palmer-Jones, R. (2001): Irrigation Service Markets in Bangladesh: Private Provision of Local Public Goods and Community Regulation? Paper prepared for the Workshop on Managing Common Resources at the Department of Sociology, Lund University, Sweden.

Rashid, M. A. (2008): Growth Phase-wise Water Requirements of Rice, Irrigation and Water Management Division, Bangladesh Rice Research Institute (BRRI), Gazipur, Dhaka, Bangladesh.

Rogers, P., R. Bhatia and A. Huber (1998): Water as a Social and Economic Good: How to Put the Principle into Practice, paper prepared for the meeting of the Technical Advisory Committee of the Global Water Partnership in Namibia.

Shahabuddin, Q. (2002): Position Paper on Water Resources Development for the Sixth Five Year Plan (2002-2007), General Economics Division, Planning Commission Ministry of Planning, Government of the People's Republic of Bangladesh.

Vaidyanathan, A. (2004): Efficiency of Water Use in Agriculture, *Economic and Political Weekly*, 2989-2996.

Wadud, M. and B. White (2002): The determinants of technical inefficiency of farms in Bangladesh, *Indian Economic Review*, *37*, 183-197.

World Bank (2000): Bangladesh: Climate Change and Sustainable Development, Bangladesh Report No. 21104-BD, Rural Development Unit, South Asia Region, Document of the World Bank.

World Bank (2006): Bangladesh Country Water Resources Assistance Strategy, Report No. 32312-BD, Environment and Social Development Unit, Agriculture and Rural Development Unit, Energy and Infrastructure Unit, South Asia Region, Document of the World Bank.

# Part 2

# Modelling, Monitoring and Assessment Techniques

# 7

# Modelling Current and Future Pan-European Irrigation Water Demands and Their Impact on Water Resources

Tim Aus der Beek[1], Ellen Kynast[2] and Martina Flörke[2]
[1]*Department of Geography, Heidelberg University*
[2]*Center for Environmental Systems Research, University of Kassel*
*Germany*

## 1. Introduction

70% of the gross global water abstractions from water resources can be explained by water withdrawals for irrigation purposes (Portmann et al., 2010). This number even rises to 90% when considering the global water consumption (Siebert et al., 2010), which is also called net irrigation water use. Furthermore, the need of irrigating field crops highly correlates with climatic conditions, which leads to intense irrigation applications in warm and water scarce regions. In pan-Europe this especially holds true for the semi-arid regions in Mediterranean countries, such as in Spain, Israel, and Turkey (Aus der Beek et al., 2010). Therefore, it is important to analyze the impact of these water withdrawals on existing water resources in order to evaluate the consequences for sustainable water management.

Within this model experiment first the historic and current net and gross irrigation water requirements are being spatially explicitly calculated for pan-Europe. The next step includes integrating these irrigation water uses in a hydrological model on the same spatial and temporal domain. After the successful validation and verification of the model results, both for irrigation and hydrology, by comparing them to reported national irrigation sums and observed river runoff data, the model concept is being transferred to simulate potential future changes of and global change impacts on irrigation water use for the 2050s. Hereby, the effects of climate change and socio-economic change on future irrigation withdrawals and water resources are being evaluated separately. Socio-economic impacts on irrigation water withdrawals are mainly being expressed by increasing or decreasing spatial irrigated extents. Here, several factors, amongst others increases in food demand due to increasing world population (Lutz et al., 2008), changing human dietaries (Hanjra & Qureshi, 2010), biofuel production (Timilsina & Shresta, 2011), influence these future irrigated extents. Another important factor is climate change (Schlenker & Lobell, 2010; Olesen & Bindi, 2002), as it not only is able to reduce local yields due increasing air temperature and climate variability but also to increase local yields due to high atmospheric $CO2$-concentrations (Long et al., 2006). Schaldach et al., (2011a) provide for the first time a separation of these influencing factors on future changes in irrigated areas and irrigation volumes for pan-Europe. In the here conducted study the same model set-up has been used to further analyze the impact of these factors not only on irrigation volumes but also on the consequences for pan-European water resources.

## 2. Material and methods

### 2.1 The study region

The pan-European study region has been developed within the EU-FP6 project SCENES which provides different pathways for the future of pan-European freshwater resources. It includes all European countries as well as their neighboring states, reaching from Northern Africa in South to the Near East in the South-East to the Russian Ural Mountains in the East (see Figure 1). The southern and eastern borders of the study region have been derived from river basin boundaries and are thus not concordant with political borders. An overview about the design of the study region and the contents of the SCENES project is given in Kamari et al., (2008). Based on the UN-classification the pan-European study region has been further divided into seven sub-regions to better allow for the analysis of regional differences: NA (Northern Africa), WE (Western Europe), NE (Northern Europe), SE (Southern Europe), EEc (Eastern Europe, central), EEe (Eastern Europe, eastern), and WA (Western Asia).

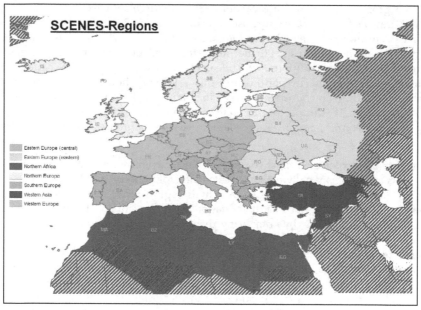

Fig. 1. Spatial extent of the study area and SCENES-regions.

### 2.2 The WaterGAP3 model

Within this integrated model study the well tested hydrology and water use model WaterGAP3 (Alcamo et al., 2003; Flörke and Alcamo, 2004; Verzano, 2009) is being applied to calculate irrigation water abstractions and their impact on pan-European water resources on a five arc minute grid (~6x9 km per grid cell). All irrigation and hydrology model runs are being conducted for the time period 1961 to 1990 (baseline) and 2041 to 2070 (scenarios 2050s). The calculation of future changes in irrigated areas based on socio-economic drivers (see chapter 2.3) is being conducted by the land use model LandSHIFT (Schaldach et al., 2011b) on the same grid. A detailed overview about the LandSHIFT model and the coupling procedure of LandSHIFT and WaterGAP within this study is given in Schaldach et al., (2011a).

First of all, based on expected population dynamics, food demand, etc. the crop-specific irrigated area maps are being generated by the LandSHIFT model and then fed into the irrigation module of WaterGAP3. The irrigation model then calculates net and gross irrigation demands as described is Aus der Beek et al., (2011):

The start day of the growing season is being calculated for each grid cell separately. For each grid cell the most suitable 150-day period within a year is ranked based on crop specific precipitation and air temperature criteria as given in Allen et al., (1998). The temperature criterion ensures continuous energy supply and optimal growing conditions, whereas the precipitation criterion promotes water supply and prevents cropping periods during droughts. If a day fulfils one of the two criteria, one ranking point is given. The growing season is then defined to be the most highly ranked 150-day period; in case of two consecutive growing periods the combination with the highest total number of ranking points is chosen (Döll & Siebert, 2002). If a second 150-day growing period is suitable, based on the crop specific precipitation and air temperature criteria, then a second cropping period within one year is added to the first period. However, in the current model set up it is not possible to change the crop type for the second period, as the crop-specific land use map provided by LandSHIFT only contains one crop type per grid cell. Therefore, double cropping is always being conducted with the same crop type, which holds true only for some crops in pan-Europe. Furthermore, the assumption of a growing period of 150 days is reasonable for crops such as vegetables, potatoes, pulses, wheat, barley, maize, rice and fruits, but underestimated for fibres and winter wheat, and overestimated for fodder plants (Smith, 1992).

Finally, the net irrigation requirements for each grid cell are being calculated, which are based on the CROPWAT approach published by Smith, (1992):

$$I_{net} = k_c * E_{pot} - P_{eff} \qquad \text{if } E_{pot} > P_{eff}$$

$$\text{(1)}$$

$$I_{net} = 0 \qquad \text{if } E_{pot} \leq P_{eff}$$

with

$I_{net}$ = net irrigation requirement per unit area [mm/d]
$P_{eff}$ = effective precipitation [mm/d]
$E_{pot}$ = potential evapotranspiration [mm/d]
$k_c$ = crop coefficient [-]

Aus der Beek et al., (2011) state further: Within the WaterGAP hydrology and water use modelling framework $E_{pot}$ is consistently being calculated accordingly to Priestley & Taylor, (1972) as a function of air temperature and net radiation (Weiß & Menzel, 2008). $K_C$ values feature a crop specific distinctive distribution curve throughout the growing period and are closely related to LAI development (Liu & Kang, 2007), as they mimic plant development. Each crop has three to four different development stages during its 150-day growing period: nursery (rice only), crop development, mid-season, and late-season.

Finally, the calculation of gross irrigation requirements $I_{gr}$ for each grid cell is being conducted by taking into account net irrigation requirements $I_{net}$ and national irrigation project efficiencies $EF_{proj}$ (Rohwer et al., 2006):

$$I_{gr} = \frac{I_{net}}{EF_{proj}} \tag{2}$$

Irrigation project efficiency reflects the state of irrigation technology within each country. It is also more applicable than the often used irrigation field efficiency as it additionally considers conveyance losses, field sizes and management practices, while irrigation field efficiency mainly results from the irrigation practice (e.g. surface, sprinkler, micro irrigation). $EF_{proj}$ typically ranges between 0.3 and 0.8, whereas 0.8 means that 80% of the water delivered to the crop is actually absorbed by it. Future changes in irrigation efficiency, have been derived by stakeholder meetings within the European research project SCENES. An overview of the performance of the WaterGAP3 irrigation module can be found in Aus der Beek et al., (2010), where its output has been compared to simulated gross irrigation requirements from a vegetation model and to reported values for all pan-European countries on a national basis.

Then, the calculated temporal and spatial explicit data sets on net irrigation requirements are being integrated in the hydrological module of WaterGAP3 (Alcamo et al., 2003; Döll et al., 2003) to assess the impact of irrigation on pan-European water resources. Here, within each irrigated grid cell they are abstracted from the internal water fluxes, and can thus alter river runoff. Also, by relating the amount of water which is being withdrawn for irrigation purposes to the amount of water that is naturally available on grid cell or river basin level, we are able to determine local water stress factors and the sustainability impact of the withdrawals. Furthermore, as WaterGAP3 also computes water withdrawals from other sectors, such as households, manufacturing industries, electricity production (Flörke et al., 2011), and livestock, we can provide an overview of locally dominant water use sectors in pan-Europe and their competition.

## 2.3 The scenarios

### 2.3.1 Socio-economic change

The SCENES project provides four different narrative socio-economic scenarios from which two opposing scenarios have been selected for this study, one reference scenario (Economy First) and one policy scenario (Sustainability Eventually). The aim of the scenarios is to provide a basis for the mid- to long-term development planning of pan-European freshwater resources. All scenarios have been designed by applying the story-and-simulation methodology (Alcamo, 2008) which iteratively links storyline revision with modeling exercises. The qualitative drivers for the scenarios have been developed in participatory international panel meetings and consider also environmental factors. The quantitative, i.e. numerical, drivers have been derived from modeling results, which are also influenced by the qualitative drivers, e.g. questionnaires filled out by panel participants (Schaldach et al., 2011a). Therefore, both scenarios offer a consistent set of environmental and socio-economic assumptions for the 2050s, which serve as a basis to study the potential future pathways of irrigation and hydrological developments in our analysis. Here, agricultural development, i.e. irrigated crop production and the impact of technological change, are the most important drivers. The two selected scenarios have been described by Schaldach et al., (2011a) as follows:

- *"Economy First"* (EcF): The economy develops towards globalisation and liberalisation, so innovations spread but income inequality, immigration and urban sprawl cause social tensions. Global demand for food and bio-fuels drives the intensification of agriculture. As the Common Agricultural Policy (CAP) is weakened, farms are abandoned where crop production is uneconomic. Until 2050 technological change allows potential increases of crop yields by 23% within the countries of the European Union (EEc, NE, WE, SE). Countries located in the other regions (EEe, NA, WA) only achieve a 14% potential increase. Total crop production is growing by 29% (from 981.890 kt to 1.266.157 kt). NA has the largest increase (+155%) followed by WA (+88%) and NE (+20%). Only for EEc a decrease of crop production by -4% is assumed. Future trends in population and economic activity show a further increase of population by 32.5% (348 million people) for pan-Europe until 2050. Here, highest growth rates are expected in NA and WA while the population increase in Europe is rather moderate. Economic activity continues to grow over the whole scenario period resulting in an 86% growth in GDP.
- *"Sustainability Eventually"* (SuE): Europe transforms from a globalised, market-oriented to an environmentally sustainable society, where local initiatives are leading. Landscape is the basic unit and there is a strong focus on quality of life. Direct agriculture subsidies are phased out and replaced by policies aimed at environmental services by farmers, such as support for farmers in less favourable areas with high-nature value farmland and accompanied by effective spatial decentralisation policies. Land use changes in general promote greater biological diversity. Crop yields are assumed to potentially increase by 50% until 2050 in all regions. Total crop production is increasing by 6.9 % (from 981.890 kt to 1.049.608 kt) with large regional differences. While crop production is doubling in NA, there is a decrease of --21% in EEe. Population is expected to increase by 13% (143 million people) in pan-Europe between 2000 and 2050. For Europe, a decrease in population is projected whereas for NA and WA the population continues to grow. Compared to EcF, SuE shows a lower total GDP development indicated by developing slower with an increase of 14% between 2000 and 2050.

## 2.3.2 Climate change

Both climate change scenarios are based on the A2 emission scenario of the IPCC SRES 4th assessment report and are combined in this study with both socio-economic scenarios. Within the A2 scenario the atmospheric $CO_2$-concentration rises up to 492 ppm (IPCC, 2007). In order to include the variability of climate models, which are being employed to calculate climate data sets with the input from the IPCC SRES report, climate output for the A2 scenario from two diverging General Circulation Models (GCMs) have been selected for this study.

The MIMR GCM output has been provided by the MICRO3.2 model at the Center for Climate System Research at the University of Tokio, Japan. Here, the A2 scenario projects high air temperature increases over Europe in combination with low precipitation decreases to high precipitation increases. The MIMR climate data set can be considered as the "wetter" scenario of the two GCMs.

The IPCM4 GCM output has been generated by the IPSL-CM4 model at the Institute Pierre Simon Laplace in France. Here, the A2 scenario indicates higher air temperature increases for Europe than the MIMR GCM and only small changes in precipitation patterns. Thus, within this study the IPCM4 climate data set can be regarded the "dry" scenario of the two GCMs.

The GCM outputs have been downscaled from the original resolution of a T63 grid (1.875°x1.875°) to the 5′ grid of the WaterGAP3 model by applying a bilinear interpolation algorithm. Furthermore, the delta change approach (Henrichs & Kaspar, 2001) has been applied to scale the GCM model output with the observed climate data for the climate normal period (1961 – 1990) from CRU (see chapter 2.2) to include climate variability in this analysis.

## 3. Results and discussion

### 3.1 Irrigation

As this study focuses on the impact of irrigation on available water resources, we refer to Schaldach et al., (2011a) for a detailed description of changes in land use patterns and irrigated area extents.

A spatial overview of mean annual pan-European net irrigation water requirements for the baseline period is given in Figure 2. The water demand is highest in the Mediterranean countries, especially in Turkey, Spain, and Italy, which account with 92040 km² for about 55% of the total real irrigated area in Europe (Aus der Beek et al., 2010). Also, the riparian zones of the Nile River as well as its Delta are heavily irrigated, both in area and quantity, which can be explained by the concurring semi-arid to arid climate conditions and population pressures. The quantitative summary for each region is given in Table 1. As explained earlier the regions surrounding the Mediterranean Sea features the highest demands: Southern Europe (16.15 bil m³), Northern Africa (15.6 bil m³), Western Asia (13.44 bil m³), followed by Eastern Europe, eastern (5.47 bil m³), Western Europe (2.38 bil m³), Northern Europe (0.47 bil m³), and Eastern Europe, central (0.36 bil m³). A country based evaluation of the goodness of these model results is given in Aus der Beek et al., (2010), who show that the deviation between modelled and reported irrigation requirements for Europe is about 1%.

Table 1 also summarizes the mean net irrigation requirements for the eight scenario model runs conducted within this study (see Chapter 2.3). The first two scenario model runs have been driven with the A2 model output from two GCMs, IPCM4 and MIMR. The socio-economic drivers, here summarized as land use, have remained in baseline conditions in order to solely analyze the impact of climate change on net irrigation water demands. Both scenarios lead to a small decrease in water demand (-1% and -5%), which is unexpected, as increasing air temperatures naturally cause an increase in evapotranspiration for most crops. Here, the decrease in irrigation water demands originates from the model structure which features a dynamic cell-specific cropping calendar. Based on the climate conditions of each modelled year the most suitable 150 day growing period, and thus the sowing day, is chosen. Therefore, changing climatic conditions shift the sowing dates to earlier or later periods to avoid high irrigation demands in July and August. A more detailed description as well as a graphic example for the Iberian Peninsula of this model algorithm can be found in Schaldach et al., (2011a). In general, this algorithm has rightly been implemented to mimic sowing date decisions from local farmers who would in reality also adapt to changing conditions in order to save expenses for irrigation water and also receive high yields. A model control run with sowing dates from the baseline for the IPCM4 scenario has shown that without these adaptation measures, the net irrigation demand increase by 15% to 61.85 bil m³ instead of decreasing by 1%. An overview of the spatial distribution of all eight scenario model runs is given in Figure 3.

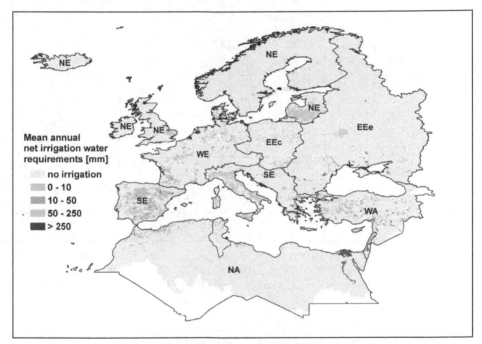

Fig. 2. Mean annual net irrigation requirements in [mm] for the baseline period (1961-90) as modeled with WaterGAP3 (EEc = Eastern Europe central, EEe = Eastern Europe east, NA = Northern Africa, NE = Northern Europe, SE = Southern Europe, WE = Western Europe, WA = Western Asia).

The next two scenario model runs have been conducted with climate input from the baseline but different socio-economic drivers, i.e. the opposing Economy First (EcF) and Sustainability Eventually (SuE) scenarios. Here, the model results show a completely different picture. Both scenarios imply an increase in net irrigation water demand, the optimistic SuE scenarios projects a minor increase of 2% for the 2050s whereas the pessimistic EcF scenario expects an increase of 48%. The differences in the spatial allocation of the water demand are also depicted in Figure 3a and 3b. Especially, in Southern Europe, namely Spain and Italy the opposing trends are evident which is also supported by Table 1 where under EcF conditions an increase of 51% and under SuE conditions an decrease of 11% occurs.

The last four scenario model runs have been conducted as combinations of the climate and socio-economic model drivers. Once again the socio-economic drivers dominate the future potential changes in irrigation water requirements. Under both "optimistic" SuE scenario model runs water demands are stable or decreasing, whereas under both "wetter" climate MIMR scenario model runs the trends are not consistent (-7% vs. +25%). As expected, the highest increase in irrigation water requirements can be observed when combing the "dry" IPCM4 scenario with the "pessimistic" Economy first scenario (+45%).

| Climate forcing | Base 61-90 | IPCM4 2050 | MIMR 2050 | Base 61-90 | Base 61-90 | IPCM42050 | IPCM4 2050 | MIMR 2050 | MIMR 2050 |
|---|---|---|---|---|---|---|---|---|---|
| Land use | Base 2000 | Base 2000 | Base 2000 | EcF 2050 | SuE 2050 | EcF 2050 | SuE 2050 | EcF 2050 | SuE 2050 |
| Eastern Europe (central) | 0.36 | 0.43 (+21) | 0.41 (+13) | 0.72 (+101) | 0.40 (+12) | 0.88 (+145) | 0.49 (+37) | 1.04 (+188) | 0.56 (+57) |
| Eastern Europe (eastern) | 5.47 | 5.93 (+8) | 5.91 (+8) | 7.39 (+35) | 3.81 (-30) | 8.12 (+49) | 4.20 (-23) | 7.32 (+34) | 3.77 (-31) |
| Northern Africa | 15.60 | 15.17 (-3) | 14.35 (-8) | 21.01 (+35) | 19.48 (+25) | 19.63 (+26) | 18.41 (+18) | 18.64 (+19) | 17.40 (+12) |
| Northern Europe | 0.47 | 0.49 (+5) | 0.46 (-2) | 1.20 (+157) | 0.46 (-2) | 1.34 (+187) | 0.51 (+8) | 1.09 (+134) | 0.40 (-15) |
| Southern Europe | 16.15 | 15.68 (-3) | 14.90 (-8) | 24.35 (+51) | 14.43 (-11) | 22.77 (+41) | 13.97 (-14) | 14.91 (-8) | 11.49 (-29) |
| Western Asia | 13.44 | 12.59 (-6) | 12.83 (-9) | 16.79 (+25) | 12.50 (-7) | 15.53 (+16) | 11.66 (-13) | 15.40 (+15) | 11.53 (-14) |
| Western Europe | 2.38 | 2.77 (+16) | 2.37 (-0) | 8.42 (+253) | 4.03 (+69) | 9.79 (+311) | 4.69 (+97) | 8.67 (+264) | 4.82 (+103) |
| SUM | 53.87 | 53.06 (-1) | 51.23 (-5) | 79.87 (+48) | 55.11 (+2) | 78.06 (+45) | 53.92 (+0) | 67.08 (+25) | 49.98 (-7) |

Table 1. Modeled mean net irrigation water requirements in billion m³ under the IPCC-SRES scenario A2 with two GCMs (IPCM4 and MIMR) for 2040 – 2069 and two socio-economic land-use scenarios (Economy First (EcF) and Sustainability Eventually (SuE)). Numbers in parenthesis describe relative changes compared to the baseline (1961 – 1990), expressed in percent.

However, as the decreasing influence of the climate drivers also lowers the combined water demand due to the adaptive sowing date, it does not top the model run with baseline climate drivers and the EcF scenario (+48%), which can be regarded the worst case scenario. A graphic overview of the combined model run outputs is given in Figure 3e to 3h.

## 3.2 Hydrological impacts

Within this study the focus has been set on modelling irrigation water withdrawals and their impact on pan-European water resources. However, in order to reach this goal we also need to analyze and quantify the competition of the irrigation water use sector with other sectors. As WaterGAP3 is a state-of-the-art model it additionally considers the other water use sectors: households, electricity generation, and manufacturing industries (see Chapter 2.2). Thus, we have calculated all sectoral water uses on river basin level and ranked their impact for each basin separately. Figure 4 features a pan-European map with dominant water use sectors, where several trends are evident. In the majority of North European river basins the manufacturing sector, e.g. in Scandinavia, and the domestic sector, e.g. in the United Kingdom and Iceland, dominate the water uses. Only Denmark is an exception, as irrigation heads the ranking here. Western and Eastern Europe feature the electricity generation sector as the most used water sector. Southern Europe, Northern Africa, and Western Asia show the irrigation sector to be the most important water use sector due to unfavourable climatic conditions and high population pressures. The patchy composition of water use sectors in Northern Africa, i.e. Libya and Algeria, originates from the location of irrigable areas.

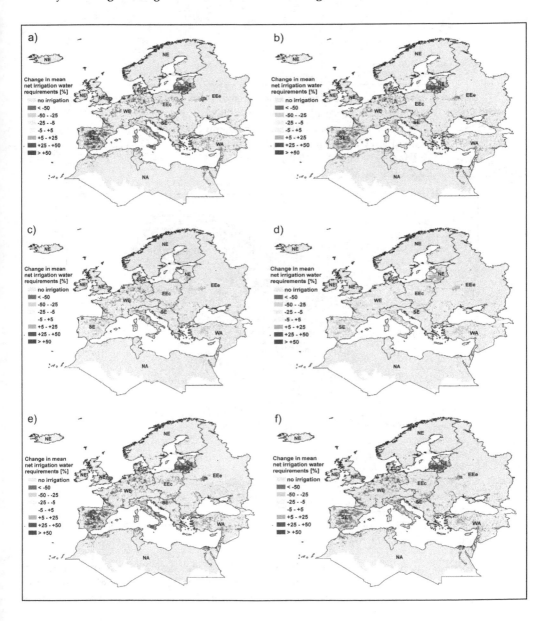

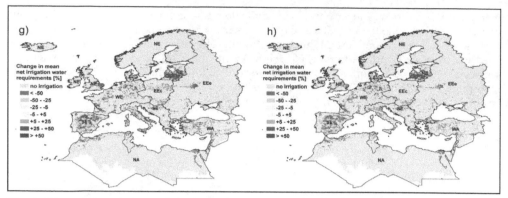

Fig. 3. Change in mean annual net irrigation requirements for the 2050s compared to baseline (1961 -1990) for different combinations of climate (CLIM: baseline; IPCM4; MIMR) and agricultural (AG: baseline; Economy First; Sustainability Eventually) scenarios. a) CLIM: baseline, AG: EcF; b) CLIM: baseline, AG: SuE; c) CLIM: IPCM4, AG: baseline; d) CLIM: MIMR, AG: baseline; e) CLIM: IPCM4, AG: EcF; f) CLIM: IPCM4, AG: SuE; g) CLIM: MIMR, AG: EcF; h) CLIM: MIMR, AG: SuE.

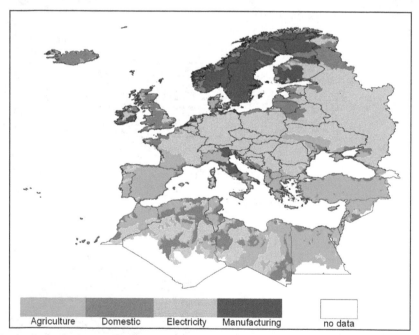

Fig. 4. Most dominant type of water use for the year 2005 on river basin level as modeled with WaterGAP3 for pan-Europe.

It needs to be mentioned that Figure 4 is an example for the year 2005. As our modelling runs are starting in 1961, the spatial as well as numerical patterns are changing within the modelling period. These patterns are influenced by a multifold of model drivers, such as climate, population numbers and allocation, power plant types and location, gross domestic product, etc., which all change with time.

To analyze the importance of irrigation water abstractions for current but also for future conditions we have summarized the shares of all water use sectors on a regional basis. The results for the year 2005 as well as exemplarily for one scenario combination are shown in Figure 5. The description of the spatial distribution of dominant water use sectors, as explained above for Figure 4, is well depicted in Figure 5a. Western Asia, Southern Europe, and Northern Africa feature with about 50% to 70% the largest irrigation water use share, followed by water used for electricity generation. Western and Eastern Europe use about 40% to 60% of their total water withdrawals for electricity generation, followed by households and manufacturing industries. Northern Europe features with about 35% equal shares for households and electricity generation. The results for the 2050s scenario driven with output from the IPCM4 climate model and socio-economic data from the Economy First set, is given in Figure 5b. Here, irrigation water withdrawals in the Mediterranean countries decrease from 50-70% in 2005 to 35-60%, whereas they remain stable or even slightly increase in the other pan-European regions. As climate change does not significantly affect future irrigation water requirements due to the adaptation measures (see Chapter 3.1), the decrease can be derived from two main factors. Firstly, the technical development of irrigation machinery has led to an improvement of the net-to-gross irrigation efficiency ratio (see Equation 2 in Chapter 2.2), reducing irrigation water withdrawals. For example, in Greece the efficiency increased from 0.57 to 0.65 and in Italy from 0.72 to 0.8. Secondly, population numbers for this scenario are decreasing in pan-Europe, except for Northern Africa and Western Asia. This trend is also well shown in Figure 5a and 5b, as water use shares in the household sector are consistently decreasing in mainland Europe. In all pan-European regions the electricity generation water use sector gains shares or its share remains constant, as for example in Western Europe. Similar patterns can be observed for the manufacturing industries water use sector.

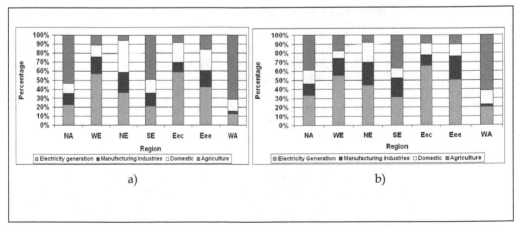

Fig. 5. Relative share of different water use sectors for pan-European regions as modeled with WaterGAP3: a) year 2005; b) scenario 2050s Economy First/IPCM4 (Eec = Eastern Europe central, Eee = Eastern Europe east, NA = Northern Africa, NE = Northern Europe, SE = Southern Europe, WE = Western Europe, WA = Western Asia).

The next step in this study is the assessment of the impacts of the irrigation water withdrawals explained above on pan-European water resources. Therefore, we have analyzed for all pan-European river basins how irrigation water abstractions affect water stress. As especially during summer time the irrigation water demand is highest, and an annual average stress indicator would mask seasonal streamflow variability, we have analyzed summer water stress induced by irrigation water abstractions, which is displayed in Figure 6. The irrigation WTA (withdrawal-to-availability ratio) indicator is a simple but effective tool to analyze water stress (Cosgrove & Rijsberman, 2000) as it divides irrigation water withdrawals by water availability on river basin level. If less than 20% of the available water resources in a river basin are being exploited, the status of this basin can be defined as low water stress and the abstractions in terms of water quantity can considered as sustainable. Medium water stress is occurring when WTA is between 20% and 40%. If more than 40% of the available water resources within a river basin are being abstracted from the system, the basin endures high water stress and the withdrawals can be considered as unsustainable. A high WTA also affects ecosystem services as environmental flow thresholds, which ensure water limits for flora and fauna, are often not being abode. Figure 6 shows the mean summer irrigation WTA for baseline conditions as well as for the two opposing scenarios combinations IPCM4/Economy Firs and MIMR/ Sustainability Eventually. The baseline results feature high irrigation induced summer water stress in Spain, Turkey, Israel, Greece, Morocco, Libya, and Algeria. Medium stressed river basins are located in Italy (e.g. Po River basin), France, and Morocco. Generally, these countries can be divided into two classes. First, semi-arid to arid regions which suffer low water availability due to climatic conditions, where already small water abstractions drastically increase water stress and water scarcity, as for example in Northern Africa and Western Asia. Secondly, semi-arid to humid regions which overexploit existing water resources, e.g. in Spain and Italy.

The "pessimistic" scenario combination IPCM4/Economy First for the 2050s increase irrigation induced summer water stress in several regions in pan-Europe (see Figure 6b). Here, most parts of France endure high water stress, except for the Rhone River basin, which experiences medium water stress. Also, in the Dniester River basin in the Ukraine water stress increases, as well as in Morocco, Algeria, Portugal, Italy, Germany, Sweden, and the United Kingdom.

The "optimistic" scenario combination MIMR/Sustainability Eventually yields an indifferent picture of pan-European summer water stress (see Figure 6c). In some river basins the water stress level decreases, as for example in Spain, Italy, and Ukraine, whereas other basins experience an increase in water stress, for example in France and Morocco. The reasons for these changes can be found in the high spatial variability of the climate change scenario data and the regional differences in the quantification of the model drivers of the socio-economic scenarios.

The next step in this study includes the impact analysis of water withdrawals from all water use sectors on water availability in pan-Europe. Therefore, we have calculated the mean annual water availability for baseline conditions after subtracting water uses, which is displayed in Figure 7a. High water availability of more than 300 mm per year occurs in Northern Europe, in the alpine basins including the Rhine, as well as in the Balkan Mountains. Medium water availability of 100 mm to 300 mm can be observed in Eastern Europe from Poland to Russia, in Spain, as well as in the countries adjacent to the Black Sea. Low water availability of less than 100 mm per year can be found in Northern Africa.

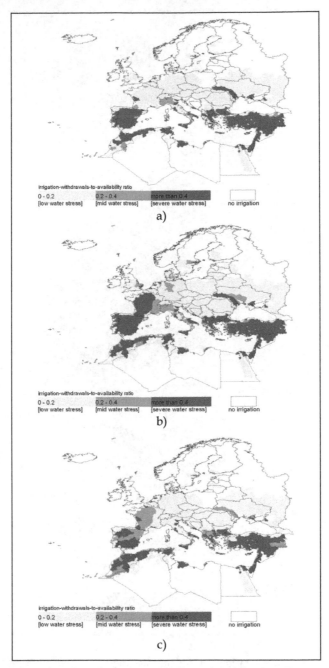

Fig. 6. Irrigation water stress in summer on river basin level for pan-Europe as modeled with WaterGAP3: a) baseline; b) scenario IPCM4/Economy First; c) scenario MIMR/Sustainability Eventually.

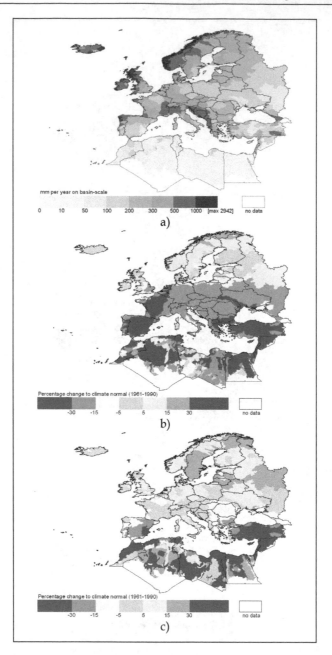

Fig. 7. a) Mean annual water availability for the baseline on river basin level for pan-Europe as modeled with WaterGAP3 with water uses; Relative change of mean annual water availability for the 2050s: b) scenario IPCM4/Economy First; c) scenario MIMR/Sustainability Eventually.

Figure 7b depicts the relative changes in water availability under the "pessimistic" IPCM4/Economy First scenario combination. High decreases of more than 30% can be observed in large parts of Spain, France, Turkey, and Israel, which is concordant with the findings of the irrigation induced summer water stress analysis (see Figure 6b). This leads to the conclusion, which is also supported by the data analysis that large parts of these decreases can be ascribed to irrigation water uses. Decreases of 15% to 30% are occurring in Central und Eastern Europe, whereas large parts of Northern Europe feature increases of 5% to 30%. The patchy patterns in Northern Africa can be explained with generally low water availabilities in the region, leading to large positive and negative changes in local water availabilities.

Figure 7c depicts the relative changes in water availability under the "optimistic" MIMR/Sustainability Eventually scenario combination. Here, decreases larger than 15% only occur in Western Asia, Northern Africa, and Spain. Also, in contrast to the pessimistic scenario described above, Central and Eastern Europe, except for Hungary, feature stable to increasing trends in water availability.

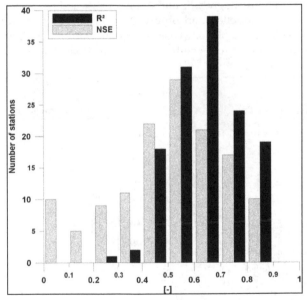

Fig. 8. Evaluation of the WaterGAP3 model performance: comparison of modelled and observed river runoff for 134 gauging stations in pan-Europe ($R^2$: coefficient of determination; NSE: Nash-Sutcliffe efficiency).

In order to analyze the performance of the hydrological module of the WaterGAP3 model, and thus the plausibility of the results of this study, we have compared observed to modelled river runoff at all river gauging stations available at Global Runoff Data Center (GRDC, 2004). Totally, runoff data from 152 stations have been available for the pan-European extent and the temporal domain of this study, whereas 18 stations have been deleted due to unrealistic data assumptions and trend tests. The goodness of fit between observed and modelled data has been evaluated by calculating the coefficient of determination $R^2$ and the Nash-Sutcliffe efficiency NSE (Krause et al., 2005) for each station. A histogram of the distribution of $R^2$ and NSE is given in Figure 8. Generally, an average $R^2$ of 0.64 with minimum and maximum values of 0.25 and 0.87 have been calculated. The NSE parameter, which is more sensitive to deviations in peak flows, features an average value of 0.5, whereas minimum and maximum values span a range of 0.01 to 0.86. The sensitivity of NSE is apparent when analyzing Figure 8. 35 stations, which is about 25% of the total station number, have a NSE smaller than 0.4, whereas only 3 stations (2%) feature a $R^2$ smaller than 0.4. This leads to the conclusion that at these 35 stations the magnitude of peak flows could not be very well represented by WaterGAP3, which is also supported by the visual analysis of the hydrographs. To display the differences in both evaluation criteria we have selected two out of the 134 hydrographs; one where the difference is large (Figure 9a) and one with small differences (Figure 9b). Figure 9a features a hydrograph of the Italian Adige River at the gauging station Trento which has a river basin size of 10049 km². The overestimation of peak flows, for example in summer 1967, as well as the underestimation of base flows, e.g. in winter 1969, leads to a poor NSE criterion of 0.14. However, as timing of peak and base flow is generally well represented by WaterGAP3, and overall volume errors balance out, the $R^2$ criterion shows a high value of 0.71. The opposite case where both, magnitude and timing of base and peak flows, are synchronic is given in Figure 9b. Here, modelled and observed runoff of the Duero River at the Spanish gauging station Villachica is displayed (basin size: 40513 km²). The high agreement of both data sets is reflected in the high model performance of 0.86 for both criteria, $R^2$ and NSE.

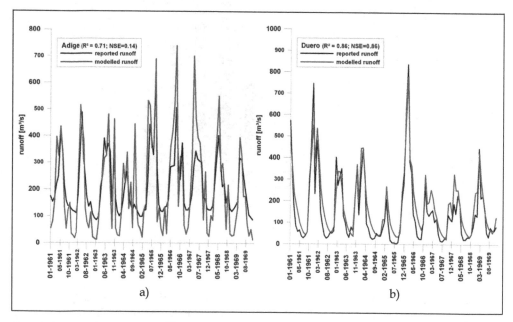

Fig. 9. Observed and modelled river runoff for 1961 to 1970: a) Adige River at station Trento (basin size 10049 km²); b) Duero River at station Villachica (basin size 40513 km²).

An analysis of the performance of the irrigation module in WaterGAP3 has been carried out in Aus der Beek et al. (2010, 2011).

## 3. Conclusions

Within this study the pan-European irrigation requirements as well as their impact on water resources has been analyzed and quantified by applying the continental hydrology and water use model WaterGAP3. Three regional hot spots of excessive irrigation water use have been identified which are all located in the vicinity of the Mediterranean Sea: Southern Europe, Western Asia, and Northern Africa. For the baseline period 1961 to 1990 about 84% of all pan-European irrigation water withdrawals occur in these three regions. Here, in opposition to the other regions, irrigation is also the dominant water use sector, except for Denmark and parts of the Ukraine. In Western and Eastern Europe water use for electricity generation is the largest sector, whereas it is domestic water use in the United Kingdom, and water use for manufacturing industries in Scandinavia. High unsustainable irrigation water withdrawals, especially in the often semi-arid and water scarce Mediterranean rim countries, lead to summer water stress, as mostly irrigation occurs in the dry and hot summer months. The water-stressed Mediterranean river basins can be separated into two classes: a) generally water scarce basins due to unfavourable climatic conditions, where already small water withdrawals drastically increase water stress (i.e. in Northern Africa and parts of Western Asia); b) overexploitation of water resources in semi-arid to humid regions, where sustainable irrigation applications would be possible (i.e. in large parts of Southern Europe). These model results have successfully been verified by comparing them to observed data, which has proven the plausibility of the methods applied in this study. Thus, it could be considered methodologically sound to transfer the WaterGAP3 model algorithms to calculate future scenarios of irrigation water use and their hydrological impacts. Here, the differentiation between climate change and socio-economic effects on irrigation water use has shown that model drivers such as land use change, due to changes in food demand, feature the largest impact on irrigation and thus hydrological quantities. Especially, as adaptive measures, such as shifts in crop sowing dates due to the elongated vegetation period, are already integrated in this study set-up and have shown to save 15% water for irrigation purposes. Thus, climate change impacts alone have nearly no impacts on future irrigation water requirements in this study. On the other hand, socio-economic impacts span a wide range of potential consequences for future irrigation water use of +2% to +48% for the 2050s. In combination with the smaller, often even slightly negative changes of climate change impacts, this range changes to -7% to +45%. According to the model results the dominance of the irrigation water use sector is also decreasing, as the electricity generation water use sector is gaining shares in the 2050s. In terms of future changes in summer water stress induced by irrigation water use, large differences between the best and worst case scenario can be observed. Here, especially river basins in Western and Southern Europe as well as in Northern Africa are affected, as a multifold of these basins stands at the crossroads of suffering (additional) water stress. Even more apparent differences between both scenarios are the changes in mean annual water availability. Here, Western and Eastern Europe show the most significant range, as water availability can drop by 15% to 30% in the worst case scenario, or can increase up to 15% in the best case scenario.

## 4. Acknowledgments

The authors would like to thank the Global Runoff Data Center in Koblenz, Germany for providing observed river runoff data for the stations applied in this study. Also, Christof

Schneider has contributed to this publication by generating pan-European maps. This study has been funded by the European Commission project SCENES.

# 5. References

Alcamo, J., Döll, P., Henrichs, T., Kaspar, F., Lehner, B., Rösch, T. & Siebert, S. (2003). Development and testing of the WaterGAP2 global model of water use and availability; Hydr. Sc., 48 (3): 317-337.

Alcamo, J., 2008. Environmental Futures, 2. The Practice of Environmental Scenario Analysis. Elsevier B.V., New York, USA.

Allen, R.G., Pereira, L.S., Raes, D. & Smith, M. (1998). FAO Irrigation and drainage paper 56: Crop evapotranspiration - Guidelines for computing crop water requirements. FAO - Food and Agriculture Organization of the United Nations, Rome, Italy.

Aus der Beek, T., Flörke, M., Lapola, D.M., Schaldach, R., Voß, F. & Teichert, E. (2010). Modelling historical and current irrigation water demand on the continental scale: Europe. Advances in Geosciences, 27: 79-85.

Aus der Beek, T., Voß, F. & Flörke, M. (2011). Modelling the impact of global change on the hydrological system of the Aral Sea basin. Physics and Chemistry of the Earth, 36: 684-694.

Cosgrove, W. & Rijsberman, F. (2000). World Water Vision: Making Water Everybody's Business. World Water Council, Earthscan Publications, London.

Döll, P. & Siebert S. (2002). Global Modelling of Irrigation Water Requirements. Water Resources Research, 38 (4): 8.1-8.10.

Döll, P., Kaspar, F. & Lehner, B. (2003). A global hydrological model for deriving water availability indicators: model tuning and validation. J. Hydrol., Vol. 270, 105-134.

Flörke, M. & Alcamo, J. (2004). European outlook on water use. Technical Report prepared for the European Environment Agency. Kongens Nytorv. 6. DK-1050. Copenhagen, DK URL: // http: scenarios.ewindows.eu.org/reports/fol949029

Flörke, M., Bärlund, I. & Teichert, E. (2010). Future changes of freshwater needs in European power plants. Management of Environmental Quality. 22(1): 89-104.

GRDC – Global Runoff Data Center (2004). Long Term Mean Monthly Discharges and Annual Characteristics of selected GRDC Stations. The Global Runoff Data Center: Koblenz, Germany.

Hanjra, M.A. & Qureshi, M.E. (2010). Global water crisis and future food security in an era of climate change. Food Policy, 35: 365-377.

Henrichs, T. & Kaspar, F. (2001). Baseline-A: A reference scenario of global change. In: Lehner, B., T. Henrichs, P. Döll, J. Alcamo (Eds.), EuroWasser: Model-based assessment of European water resources and hydrology in the face of global change. Center for Environmental Systems Research, University of Kassel, Kassel World Water Series – Report No. 5: 4.1-4.8.

Intergovernmental Panel on Climate Change (IPCC), Alley, R.B., Allison, I., Carrasco, J., Falto, G., Fujii, Y., Kaser, G., Mote, P., Thomas, R.H. & Zhang, T. (2007). Climate Change 2007: The Physical Science Basis. Contribution of Working Group I to the

Fourth Assessment Report of the Intergovernmental Panel on Climate Change, Cambridge Univ. Press, Cambridge, U.K.

Kamari, J., Alcamo, J., Barlund, I., Duel, H., Farquharson, F., Florke, M., Fry, M., Houghton-Carr, H., Kabat, P., Kaljonen, M., Kok, K., Meijer, K.S., Rekolainen, S., Sendzimir, J., Varjopuro, R. & Villars, N. (2008). Envisioning the future of water in Europe - the SCENES project. *E-WAter*, 2008: 1-28.

Krause, P., Doyle, D.P & Bäse, F. (2005). Comparison of different efficiency criteria for hydrological model assessment. *Advances in Geosciences*, 5, 89-97.

Liu, H.J. & Kang, Y. (2007). Sprinkler irrigation scheduling of winter wheat in the North China Plain using a 20 cm standard pan. *Irrigation Science*, 25 (2): 149-159.

Long, S.P., Ainsworth, E.A., Leakey, A.D.B., Nösberger, J. & Ort, D.R. (2006). Food for Thought: Lower-Than-Expected Crop Yield Stimulation with Rising $CO_2$ Concentrations, *Science*, 312 (5782): 1918-1921.

Lutz, W., Sanderson, W. & Scherbov, S. (2008). The coming acceleration of global population ageing. *Nature* 451: 716-719.

Olesen, J.E. & Bindi, M. (2002). Consequences of climate change for European agricultural productivity, land use and policy. *European Journal of Agronomy*, 16(4): 239-262.

Portmann, F.T., Siebert, S. & Döll, P. (2010). MIRCA2000–Global monthly irrigated and rainfed crop areas around the year 2000: A new high resolution data set for agricultural and hydrological modeling. *Global Biogeochemical Cycles*, 24, GB1011.

Priestley, C.H.B. & Taylor, R.J. (1972). On the assessment of surface heat flux and evaporation using large scale parameters. *Monthly Weather Review*, 100: 81-92.

Rohwer, J., Gerten, D. & Lucht, W. (2006). *Development of functional irrigation types for improved global crop modeling*; PIK-Report 106; PIK; Potsdam, Germany.

Schaldach, R., Koch, J., Aus der Beek, T., Kynast, E. & Flörke, M. (2011). Current and future irrigation water requirements in pan-Europe: A comparative analysis of influencing factors. *Global and Planetary Change* (in review).

Schaldach, R., Alcamo, J., Koch, J., Kölking, C., Lapola, D.M., Schüngel, J. & Priess, J.A. (2011b). An integrated approach to modelling land use change on continental and global scales. *Environmental Modelling and Software*, 26 (8): 1041-1051.

Schlenker, W. & Lobell, D.B. (2010). Robust negative impacts of climate change on African agriculture. *Env. Res. Lett.* 5, doi10.1088/1748-9326/5/1/014010.

Siebert, S., Burke, J., Faures, J.M., Frenken, K., Hoogeveen, J., Döll, P. & Portmann, F.T. (2010). Groundwater use for irrigation - a global inventory. *Hydrol. Earth Syst. Sci.*, Vol. 14, 1863-1880, doi:10.5194/hess-14-1863-2010.

Smith, M. (1992). *Irrigation and Drainage Pap. 46*, CROPWAT - A computer program for irrigation planning and management, Food and Agric. Org. of the U. N., Rome, Italy.

Timilsina, G.R. & Shrestha, A. (2011). How much hope should we have for biofuels? *Energy*, Volume 36, Issue 4, Pages 2055-2069.

Verzano, K. (2009). *Climate Change Impacts on Flood Related Hydrological Processes: Further Development and Application of a Global Scale Hydrological Model.* Dissertation University of Kassel, Germany. Available at:

www.mpimet.mpg.de/fileadmin/publikationen/Reports/WEB_BzE_71_verzano.
    pdf
Weiß, M. & Menzel, L. (2008). A global comparison of four potential evapotranspiration
    equations and their relevance to stream flow modelling in semi-arid environments.
    *Adv. Geosci.*,18, 15-23.

# 8

# Basics and Application of Ground-Penetrating Radar as a Tool for Monitoring Irrigation Process

Kazunori Takahashi[1], Jan Igel[1], Holger Preetz[1] and Seiichiro Kuroda[2]
[1]*Leibniz Institute for Applied Geophysics*
[2]*National Institute for Rural Engineering*
[1]*Germany*
[2]*Japan*

## 1. Introduction

Ground-penetrating radar (GPR) is a geophysical method that employs an electromagnetic technique. The method transmits and receives radio waves to probe the subsurface. One of the earliest successful applications was measuring ice thickness on polar ice sheets in 1960s (Knödel et al., 2007). Since then, there have been rapid developments in hardware, measurement and analysis techniques, and the method has been extensively used in many applications, such as archaeology, civil engineering, forensics, geology and utilities detection (Daniels, 2004).

There are a variety of methods to measure soil water content. The traditional method is to dry samples from the field and compare the weights of the samples before and after drying. This method can analyse sampled soils in detail and the results may be accurate. A classical instrument for in situ measurements is the tensiometer, which measures soil water tension. These methods have the disadvantages of being destructive and time-intensive, and thus it is impossible to capture rapid temporal changes. Therefore, a number of sophisticated physical methods have been developed for non-destructive in situ measurements. One of these methods is time domain reflectometry (TDR), which has been widely used for determining soil moisture since the 1980s (Topp et al., 1980; Noborio, 2001; Robinson et al., 2003). TDR measurements are easy to carry out and cost effective, however they are not suitable for obtaining the high-resolution soil water distribution because either a large number of probes have to be installed or a single measurement has to be repeated at various locations. In addition, TDR measurements are invasive; the probes must be installed into soil, which may slightly alter the soil properties. GPR has the potential to overcome these problems and is considered one of the most suitable methods for monitoring soil water content during and after irrigation because of the following features:

- The GPR response reflects the dielectric properties of soil that are closely related to its water content.
- GPR data acquisition is fast compared to other geophysical methods. This feature enables measurements to be made quickly and repeatedly, yielding high temporal resolution monitoring. This is very important for capturing rapid changes.

- GPR can be used as a completely non-invasive method. The antennas do not have to touch the ground and thus it does not disturb the natural soil conditions.
- GPR systems are compact and easy to use compared to other geophysical methods. This feature enables scanning over a wide area and the collection of 2D or 3D data. Further, the distribution of soil properties can be obtained with high spatial resolution.

The objective of this chapter is to provide the basics of GPR and examples of its application. Readers who are interested in this measurement technique can find more detailed and useful information in the references listed at the end of the chapter.

## 2. Basic principles of GPR

A GPR system consists of a few components, as shown in Fig. 1, that emit an electromagnetic wave into the ground and receive the response. If there is a change in electric properties in the ground or if there is an anomaly that has different electric properties than the surrounding media, a part of the electromagnetic wave is reflected back to the receiver. The system scans the ground to collect the data at various locations. Then a GPR profile can be constructed by plotting the amplitude of the received signals as a function of time and position, representing a vertical slice of the subsurface, as shown in Fig. 2. The time axis can be converted to depth by assuming a velocity for the electromagnetic wave in the subsurface soil.

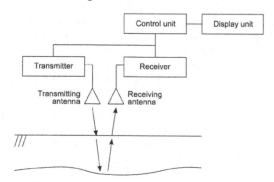

Fig. 1. Block diagram of a GPR system.

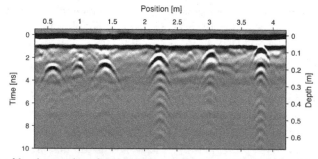

Fig. 2. A GPR profile obtained with a 1.5 GHz system scanned over six objects buried in sandy soil. The signal amplitude is plotted as a function of time (or depth) and position. Relatively small objects are recognised by hyperbolic-shaped reflections. Reflections from the ground surface appear as stripes at the top of the figure.

## 2.1 Electromagnetic principles of GPR

### 2.1.1 Electromagnetic wave propagation in soil

The propagation velocity $v$ of the electromagnetic wave in soil is characterised by the dielectric permittivity $\varepsilon$ and magnetic permeability $\mu$ of the medium:

$$v = \frac{1}{\sqrt{\varepsilon\mu}} = \frac{1}{\sqrt{\varepsilon_0\varepsilon_r\mu_0\mu_r}}$$

(1)

where $\varepsilon_0 = 8.854 \times 10^{-12}$ F/m is the permittivity of free space, $\varepsilon_r = \varepsilon/\varepsilon_0$ is the relative permittivity (dielectric constant) of the medium, $\mu_0 = 4\pi \times 10^{-7}$ H/m is the free-space magnetic permeability, and $\mu_r = \mu/\mu_0$ is the relative magnetic permeability. In most soils, magnetic properties are negligible, yielding $\mu = \mu_0$, and Eq. 1 becomes

$$v = \frac{c}{\sqrt{\varepsilon_r}}$$

(2)

where $c = 3 \times 10^8$ m/s is the speed of light. The wavelength $\lambda$ is defined as the distance of the wave propagation in one period of oscillation and is obtained by

$$\lambda = \frac{v}{f} = \frac{2\pi}{\omega\sqrt{\varepsilon\mu}}$$

(3)

where $f$ is the frequency and $\omega = 2\pi f$ is the angular frequency.

In general, dielectric permittivity $\varepsilon$ and electric conductivity $\sigma$ are complex and can be expressed as

$$\varepsilon = \varepsilon' - j\varepsilon''$$

(4)

$$\sigma = \sigma' - j\sigma''$$

(5)

where $\varepsilon'$ is the dielectric polarisation term, $\varepsilon''$ represents the energy loss due to the polarisation lag, $\sigma'$ refers to ohmic conduction, and $\sigma''$ is related to faradaic diffusion (Knight & Endres, 2005). A complex effective permittivity expresses the total loss and storage effects of the material as a whole (Cassidy, 2009):

$$\varepsilon^e = \left(\varepsilon' + \frac{\sigma''}{\omega}\right) - j\left(\varepsilon'' + \frac{\sigma'}{\omega}\right)$$

(6)

The ratio of the imaginary and real parts of the complex permittivity is defined as tan $\delta$ (loss tangent):

$$\tan\delta = \frac{\varepsilon''}{\varepsilon'} \cong \frac{\sigma'}{\omega\varepsilon'}$$

(7)

When $\varepsilon''$ and $\sigma''$ are small, it is approximated as the right most expression. In the plane-wave solution of Maxwell's equations, the electric field $E$ of an electromagnetic wave that is travelling in z-direction is expressed as

$$E(z,t) = E_0 e^{j(\omega t - kz)}$$

(8)

where $E_0$ is the peak signal amplitude and $k = \omega\sqrt{\varepsilon\mu}$ is the wavenumber, which is complex if the medium is conductive, and it can be separated into real and imaginary parts:

$$k = \alpha + j\beta$$

(9)

The real part $\alpha$ and imaginary part $\beta$ are called the attenuation constant (Np/m) and phase constant (rad/m), respectively, and given as follows:

$$\alpha = \omega\left[\frac{\varepsilon'\mu}{2}\left(\sqrt{1 + \tan^2\delta} - 1\right)\right]^{1/2}$$

(10)

$$\beta = \omega\left[\frac{\varepsilon'\mu}{2}\left(\sqrt{1 + \tan^2\delta} + 1\right)\right]^{1/2}$$

(11)

The attenuation constant can be expressed in dB/m by $\alpha' = 8.686\alpha$. The inverse of the attenuation constant:

$$\delta = \frac{1}{\alpha}$$

(12)

is called the skin depth. It gives the depth at which the amplitude of the electric field decay is $1/e$ (~ -8.7 dB, ~ 37%). It is a useful parameter to describe how lossy the medium is. Table 1 provides the typical range of permittivity, conductivity and attenuation of various materials.

| Material | Relative permittivity | Conductivity [S/m] | Attenuation constant [dB/m] |
|---|---|---|---|
| Air | 1 | 0 | 0 |
| Freshwater | 81 | $10^{-6}$-$10^{-2}$ | 0.01 |
| Clay, dry | 2-6 | $10^{-3}$-$10^{-1}$ | 10-50 |
| Clay, wet | 5-40 | $10^{-1}$-$10^{-0}$ | 20-100 |
| Sand, dry | 2-6 | $10^{-7}$-$10^{-3}$ | 0.01-1 |
| Sand, wet | 10-30 | $10^{-3}$-$10^{-2}$ | 0.5-5 |

Table 1. Typical range of dielectric characteristics of various materials measured at 100 MHz (Daniels, 2004; Cassidy, 2009).

### 2.1.2 Reflection and transmission of waves

GPR methods usually measure reflected or scattered electromagnetic signals from changes in the electric properties of materials. The simplest scenario is a planar boundary between two media with different electric properties as shown in Fig. 3, which can be seen as a layered geologic structure.

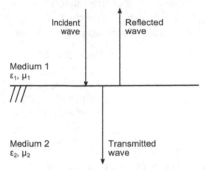

Fig. 3. Reflection and transmission of a normally incident electromagnetic wave to a planar interface between two media.

When electromagnetic waves impinge upon a planar dielectric boundary, some energy is reflected at the boundary and the remainder is transmitted into the second medium. The relationships of the incident, reflected, and transmitted electric field strengths are given by

$$E^i = E^r + E^t \tag{13}$$

$$E^r = R \cdot E^i \tag{14}$$

$$E^t = T \cdot E^i \tag{15}$$

respectively, where $R$ is the reflection coefficient and $T$ is the transmission coefficient. In the case of normal incidence, illustrated in Fig. 3, the reflection and transmission coefficients are given as

$$R = \frac{Z_2 - Z_1}{Z_2 + Z_1} \tag{16}$$

$$T = 1 - R = \frac{2Z_2}{Z_2 + Z_1} \tag{17}$$

where $Z_1$ and $Z_2$ are the intrinsic impedances of the first and second media, respectively, and $Z = \sqrt{\mu/\varepsilon}$. In a low-loss non-conducting medium, the reflection coefficient may be simplified as (Daniels, 2004)

$$R \cong \frac{\sqrt{\varepsilon_{r1}} - \sqrt{\varepsilon_{r2}}}{\sqrt{\varepsilon_{r1}} + \sqrt{\varepsilon_{r2}}} \tag{18}$$

## 2.2 GPR systems

A GPR system is conceptually simple and consists of four main elements: the transmitting unit, the receiving unit, the control unit and the display unit (Davis & Annan, 1989), as depicted in Fig. 1. The basic type of GPR is a time-domain system in which a transmitter generates pulsed signals and a receiver samples the returned signal over time. Another common type is a frequency-domain system in which sinusoidal waves are transmitted and received while sweeping a given frequency. The time-domain response can be obtained by an inverse Fourier transform of the frequency-domain response.

GPR systems operate over a finite frequency range that is usually selected from 1 MHz to a few GHz, depending on measurement requirements. A higher frequency range gives a narrower pulse, yielding a higher time or depth resolution (i.e., range resolution), as well as lateral resolution. On the other hand, attenuation increases with frequency, therefore high-frequency signal cannot propagate as far and the depth of detection becomes shallower. If a lower frequency is used, GPR can sample deeper, but the resolution is lower.

Antennas are essential components of GPR systems that transmit and receive electromagnetic waves. Various types of antennas are used for GPR systems, but dipole and bowtie antennas are the most common. Most systems use two antennas: one for transmitting and the other for receiving, although they can be packaged together. Some commercial GPR systems employ shielded antennas to avoid reflections from objects in the air. The antenna gain is very important in efficiently emitting and receiving the electromagnetic energy. Antennas with a high gain help improve the signal-to-noise ratio. To achieve a higher antenna gain, the size of an antenna is determined by the operating frequency. A lower operating frequency requires larger antennas. Small antennas make the system compact, but they have a low gain at lower frequencies.

## 2.3 GPR surveys

GPR surveys can be categorised into reflection and transillumination measurements (Annan, 2009). Reflection measurements commonly employ configurations called common-offset and common midpoint. If antennas are placed on the ground, there are propagation paths both in and above the ground, as shown in Fig. 4. Transillumination measurements are usually carried out using antennas installed into trenches or drilled wells.

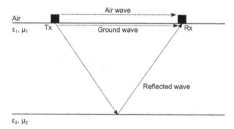

Fig. 4. Propagation paths of electromagnetic waves for a surface GPR survey with a layered structure in the subsurface.

## 2.3.1 Common-offset (CO) survey

In a common-offset survey, a transmitter and receiver are placed with a fixed spacing. The transmitter and receiver scan the survey area, keeping the spacing constant and acquiring the data at each measurement location, as depicted in Fig. 5. For a single survey line, the acquired GPR data corresponds to a 2D reflectivity map of the subsurface below the scanning line, i.e., a vertical slice (e.g., Fig. 2). By setting multiple parallel lines, 3D data can be obtained and horizontal slices and 3D maps can be constructed.

### 2.3.2 Common midpoint (CMP) survey

In a common midpoint survey, a separate transmitter and receiver are placed on the ground. The separation between the antennas is varied, keeping the centre position of the antennas constant. With varying separation and assuming a layered subsurface structure, various signal paths with the same point of reflection are obtained and the data can be used to estimate the radar signal velocity distribution versus subsurface depth (e.g., Annan, 2005; Annan, 2009). The schematic configuration of a CMP survey is shown in Fig. 5. When the transmitter is fixed, instead of being moved from the midpoint together with the receiver, and if only the receiver is moved away from the transmitter, the setup is called a wide-angle reflection and refraction (WARR) gather.

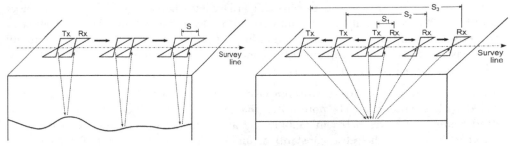

Fig. 5. Schematic illustrations of common-offset (left) and common midpoint (right) surveys. Tx and Rx indicate the transmitting and receiving antennas, respectively. Antennas are scanned with a fixed spacing $S$ in the common-offset configuration, while the spacing is varied as shown with $S_1$, $S_2$, $S_3$ ... with respect to the middle position in the common midpoint configuration.

### 2.3.3 Transillumination measurements

Zero-offset profiling (ZOP) uses a configuration where the transmitter and receiver are moved in two parallel boreholes with a constant distance (Fig. 6, left), resulting in parallel raypaths in the case of homogeneous subsurface media. This setup is a simple and quick way to locate anomalies.

Transillumination multioffset gather surveying provides the basis of tomographic imaging. The survey measures transmission signals through the volume between boreholes with varying angles (Fig. 6, right). Tomographic imaging constructed from the survey data can provide the distribution of dielectric properties of the measured volume.

### 2.4 Physical properties of soil

As seen in the previous section, the electric and magnetic properties of a medium influence the propagation and reflection of electromagnetic waves. These properties are dielectric permittivity, electric conductivity and magnetic permeability.

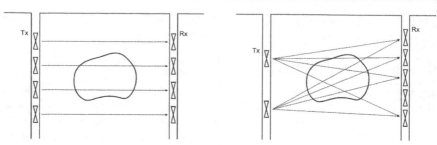

Fig. 6. Schematic illustrations of transillumination zero-offset profiling (left) and transillumination multioffset gather (right) configurations.

## 2.4.1 Dielectric permittivity

Permittivity describes the ability of a material to store and release electromagnetic energy in the form of electric charge and is classically related to the storage ability of capacitors (Cassidy, 2009). Permittivity greatly influences the electromagnetic wave propagation in terms of velocity, intrinsic impedance and reflectivity. In natural soils, dielectric permittivity might have a larger influence than electric conductivity and magnetic permeability (Lampe & Holliger, 2003; Takahashi et al., 2011).

Soil can be regarded as a three-phase composite with the soil matrix and the pore space that is filled with air and water. The pore water phase of soil can be divided into free water and bound water that is restricted in mobility by absorption to the soil matrix surface. The relative permittivity (dielectric constant) of air is 1, is between 2.7 and 10 for common minerals in soils and rocks (Ulaby et al., 1986), while water has a relative permittivity of 81, depending on the temperature and frequency. Thus, the permittivity of water-bearing soil is strongly influenced by its water content (Robinson et al., 2003). Therefore, by analysing the dielectric permittivity of soil measured or monitored with GPR, the soil water content can be investigated.

As mentioned previously, water plays an important role in determining the dielectric behaviour of soils. The frequency-dependent dielectric permittivity of water affects the permittivity of soil. Within the GPR frequency range, the frequency dependence is caused by polarisation of the dipole water molecule, which leads to relaxation. A simple model describing the relaxation is the Debye model in which the relaxation is associated with a relaxation time $\tau$ that is related to the relaxation frequency $f_{relax} = 1/(2\pi\tau)$. From this model, the real component of permittivity $\varepsilon'$ and imaginary component $\varepsilon''$ are given by

$$\varepsilon'(\omega) = \varepsilon_\infty + \frac{\varepsilon_s - \varepsilon_\infty}{1 + \omega^2 \tau^2} \tag{19}$$

$$\varepsilon''(\omega) = \omega\tau\left(\frac{\varepsilon_s - \varepsilon_\infty}{1 + \omega^2 \tau^2}\right) \tag{20}$$

where $\varepsilon_s$ is the static (DC) value of the permittivity and $\varepsilon_\infty$ is the optical or very-high-frequency value of the permittivity. Pure free water at room temperature (at 25°C) has a relaxation time $\tau = 8.27$ ps (Kaatze, 1989), which corresponds to a relaxation frequency of approximately 19 GHz. Therefore, free water losses will only start to have a significant effect with high-frequency surveys (i.e., above 500 MHz; Cassidy, 2009).

There are a number of mixing models that provide the dielectric permittivity of soil. One of the most popular models is an empirical model called Topp's equation (Topp et al., 1980), which describes the relationship between relative permittivity $\varepsilon_r$ and volumetric water content $\theta_v$ of soil:

$$\varepsilon_r = 3.03 + 9.3\theta_v + 146\theta_v^2 - 76.6\theta_v^3 \tag{21}$$

$$\theta_v = -5.3 \times 10^{-2} + 2.92 \times 10^{-2}\varepsilon_r - 5.5 \times 10^{-4}\varepsilon_r^2 + 4.3 \times 10^{-6}\varepsilon_r^3 \tag{22}$$

The model is often considered inappropriate for clays and organic-rich soils, but it agrees reasonably well for sandy/loamy soils over a wide range of water contents (5-50%) in the GPR frequency range (10 MHz-1 GHz). The model does not account for the imaginary component of permittivity.

The complex refractive index model (CRIM) is valid for a wide variety of soils. The model uses knowledge of the permittivities of a material and their fractional volume percentages, and it can be used on both the real and imaginary components of the complex permittivity. The three-phase soil can be modelled with the complex effective permittivity of water $\varepsilon_w$, gas (air) $\varepsilon_g$ and matrix $\varepsilon_m$ as (Shen et al., 1985)

$$\varepsilon^e = \left\{ \left( \phi S_w \sqrt{\varepsilon_w} \right) + \left[ (1-\phi)\sqrt{\varepsilon_m} \right] + \left[ \phi(1-S_w)\sqrt{\varepsilon_g} \right] \right\}^2 \tag{23}$$

where $\phi$ is the porosity and $S_w = \theta_v / \phi$ is the water saturation (e.g., the percentage of pore space filled with fluid). Fig. 7 shows the comparison of modelled dielectric permittivity using Topp's equation and CRIM for a sandy soil.

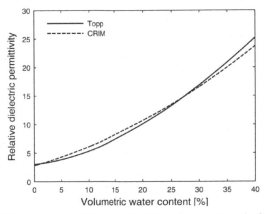

Fig. 7. The modelled dielectric permittivity using Topp's equation (solid line) and CRIM (dashed line) for sandy soil. The porosity and permittivity of the soil matrix are assumed to be $\phi = 40\%$ and $\varepsilon_m = 4.5$, respectively.

## 2.4.2 Electric conductivity

Electric conductivity describes the ability of a material to pass free electric charges under the influence of an applied field. The primary effect of conductivity on electromagnetic waves is energy loss, which is expressed as the real part of the conductivity. The imaginary part contributes to energy storage and the effect is usually much less than that of energy loss. In highly conductive materials, the electromagnetic energy is lost as heat and thus the electromagnetic waves cannot propagate as deeply. Therefore, GPR is ineffective in materials such as those under saline conditions or with high clay contents (Cassidy, 2009).

Three mechanisms of conduction determine the bulk electric conductivity. The first is electron conduction caused by the free electrons in the crystal lattice of minerals, which can often be negligible. The second is electrolytic conductivity caused by the aqueous liquid containing dissolved ions in pore spaces. The third type of conductivity is surface conductivity associated with the excess charge in the electrical double layer at the solid/fluid interface, which is typically high for clay minerals and organic soil matter. This concentration of charge provides an alternate current path and can greatly enhance the electrical conductivity of a material (Knight & Endres, 2005). The bulk conductivity of sediments and soils can be modelled as (Knödel et al., 2007)

$$\sigma = \frac{\phi^m}{a} \sigma_w S_w{}^n + \sigma_q$$

(24)

with knowledge of the conductivity of the pore fluid $\sigma_w$, effective porosity $\phi$ and water saturation $S_w$, and where $n$, $m$ and $a$ are model parameters, i.e., a saturation exponent (often $n \sim 2$), a cementation exponent ($m = 1.3$-$2.5$ for most sedimentary rocks) and an empirical parameter ($a = 0.5$-$1$ depending on the type of sediment), respectively. The first term is according to Archie's law (Archie, 1942), which represents a contribution from electrolytic conductivity. The second term adds the contribution of the interface conductivity $\sigma_q$. At GPR frequencies, the conductivity is often approximated as real-valued static or DC conductivity (Cassidy, 2009).

## 2.4.3 Magnetic permeability

The magnetic property of soils is caused by the presence of ferrimagnetic minerals, mainly magnetite, titanomagnetite, and maghemite. These minerals either stem from the parent rocks or can be formed during soil genesis.

As discussed in previous sections, the magnetic properties theoretically influence the propagation of electromagnetic waves. However, in natural soils, the influence of the magnetic properties of the soil is fairly low in most cases. The magnetic permeability must be extremely high to influence the GPR signal. For example, Cassidy (2009) suggests that magnetic susceptibility $\kappa = (\mu/\mu_0) - 1$ must be greater than $30{,}000 \times 10^{-5}$ SI to have an influence comparable to the dielectric permittivity. Soils exhibiting such high magnetic susceptibilities are extremely rare. Although tropical soils often display high susceptibilities, values in this range are exceptional (Preetz et al., 2008). Therefore, the magnetic permeability of most soils is usually assumed to be the same as that of free space, i.e., $\mu = \mu_0$ and $\mu_0 = 1$.

## 3. Soil moisture determination using GPR

There is a close relation between soil dielectric permittivity and its water content, as described in the previous section. GPR can be used to measure this proxy and the soil water content by implementing a variety of measurement and analysis techniques. They provide different sampling depths, and spatial and temporal resolutions and accuracies. Most of these techniques use a relation that links the permittivity of the soil to the propagation velocity of the electromagnetic waves (Eqs. 1 and 2). Other techniques determine the coefficient of reflection, which also depends on the dielectric permittivity of the materials (Eqs. 16 and 18).

| | Transmission, traveltime | Reflection, traveltime (CO) | Reflection, traveltime (MO) | Diffraction, traveltime (CO) | Ground wave, traveltime | Reflection coefficient |
|---|---|---|---|---|---|---|
| Depth of investigation | Some 10 m | Metres | Metres | Metres | 10 cm | A few cm |
| Accuracy | High | High | Low | Low | High | High |
| Spatial resolution | Medium - high | Medium | Low | Medium | High* | High* |
| Information on vertical moisture distribution | Yes | Yes | Yes | Yes | Limited | No |
| Cost (setup and data analysis) | High | Low | High | Medium | Low | Low |
| Requirements | Boreholes | Plane interfaces in the subsurface | Plane interfaces in the subsurface | Natural or artificial diffractors in the subsurface | Short vegetation, little surface roughness | Short vegetation, little surface roughness |

Table 2. Summary of GPR methods used to measure water content and their qualitative rating (*: only in the lateral direction).

### 3.1 Transmission measurements

One straightforward way to deduce the electromagnetic wave velocity is to measure the time that a wave takes to travel along a known distance. For example, the travel path is known for borehole-to-borehole measurements (Fig. 6), assuming a homogeneous medium and a straight path. The mean wave velocity along the travel path can be easily calculated and the mean water content can be deduced. When using a tomographic layout with several interlacing travel paths (Fig. 6, right) and after inverting the data, we can obtain a high-resolution image of the soil water distribution between the boreholes. Tomographic borehole measurements have been successfully used to assess the water content distribution and, further, hydraulic properties (e.g., Tronicke et al., 2001; Binley et al., 2001). Tomographic measurements can also be applied from one borehole to the surface or if two or more sides of the study volume are accessible, e.g., by trenches (Schmalholz et al., 2004).

## 3.2 Reflection measurements with a known subsurface geometry

When using a common-offset (CO) configuration and if the depths of radar reflectors in the subsurface are known, the wave velocity can be determined using $v = 2d/t$, where $d$ is the depth of the reflector and $t$ is the two-way traveltime. In the case of a layered subsurface, every reflection can be analysed and the interval velocity within each layer can be calculated from the mean velocities with the Dix formula (Dix, 1955):

$$v_n = \left( \frac{\bar{v}_n^2 t_n - \bar{v}_{n-1}^2 t_{n-1}}{t_n - t_{n-1}} \right)^{1/2}$$

(25)

where $v_n$ is the interval velocity of the $n$th layer, $\bar{v}_{n-1}$ and $\bar{v}_n$ are the stacking velocities from the datum to reflectors above and below the layer, and $t_{n-1}$ and $t_n$ are reflection arrival times.

The depth of reflectors can be obtained from boreholes or by digging a trench. Stoffregen et al. (2002) measured the water content of a lysimeter by analysing the traveltimes of the reflection at the bottom of the lysimeter. However in most cases, the depths of reflecting structures are not available, and the geometry of the subsurface and the electromagnetic characteristics have to be deduced from GPR measurements.

## 3.3 Multioffset measurements

If there is no knowledge about the geometry of the subsurface, further information is needed, which can be achieved by multioffset GPR measurements. Several acquisition layouts can be used to acquire multioffset data: common midpoint gather (CMP), wide angle reflection and refraction (WARR) gather (see Section 2.3.2), sequential constant offset measurements with different offsets or continuous multioffset measurements with multi channel GPR devices. After acquisition, all data can be converted to CMP sections by sorting the radar traces (e.g., Yilmaz, 2000; Greaves et al., 1996). With different antenna offsets, the waves take different propagation paths to and from a reflector in the subsurface (Fig. 5, right). A larger offset $x$ results in a longer travel path and traveltime $t$:

$$t = \frac{2\sqrt{d^2 + (x/2)^2}}{v}$$

(26)

This equation describes a hyperbola, hence horizontal reflectors are mapped as reflection hyperbolas in a CMP radar section. Fig. 8 shows a CMP section from a sandy environment where the groundwater table is at 4.6 m depth. Several reflections at boundaries within the sand are visible. The first straight onset with a slope of $1/c_0$ is the airwave, a direct wave propagating through the air from the transmitter to the receiver. The second straight onset with a steeper slope of $1/v$ is the ground wave, followed by reflection hyperbolas from several reflectors in the subsurface. By fitting hyperbolas to some of the reflections, a depth-velocity model can be constructed that gives information on the water distribution. This model can also be used to transform the data from traveltimes to depth, analogous to seismic data processing (e.g., Yilmaz, 2000). When analysing multioffset data along a profile, a 2D velocity distribution of the subsurface and, thus, the water distribution can be deduced. This technique has been used successfully to map water content in the subsurface (e.g., Greaves et al., 1996; Bradford, 2008).

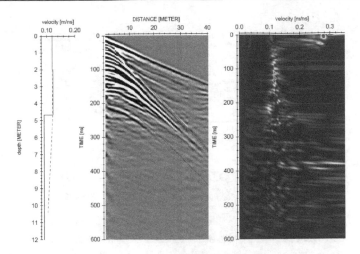

Fig. 8. Radar profile of the 80 MHz CMP measurement and fitted reflection hyperbolas (centre), semblance analysis (right) and velocity-depth model (left).

### 3.4 Diffraction measurements

Radar waves are diffracted by objects that are smaller than their wavelength, resulting in scattering in all directions. When performing a constant offset radar measurement, such objects cause a signal in the data before the antenna is directly above the object. When $x$ is the lateral distance and $d$ is the depth of the object relative to the antenna, the diffracted wave appears at

$$t = \frac{2\sqrt{d^2 + x^2}}{v_{soil}}$$

(27)

which describes a hyperbola. Objects causing such diffraction hyperbolas include stones, roots, pipes or wires if the antenna is moved perpendicular to their alignment. The mean wave velocity between the antenna and the object can be determined by fitting a synthetic hyperbola to the data and can be used to assess the mean water content above the object. This method has been used successfully to recover soil moisture distribution by analysing hyperbolas from natural objects or artificial objects that were placed in the subsurface (e.g., Igel et al., 2001; Schmalholz, 2007).

### 3.5 The ground wave

One of the most commonly used techniques for soil moisture mapping with GPR is analysing the ground wave velocity. For surface GPR measurements the ground wave is the only wave that travels through the ground with a propagation path that is known a priori, and the wave velocity can be calculated directly from the traveltime. Analysing the ground wave has proven to be a fast technique that can be used to map large areas and yield reasonable results in comparison to other methods, such as TDR or gravimetric soil moisture determination (Du, 1996; Grote et al., 2003; Huisman et al., 2001, 2003; Overmeeren et al., 1997).

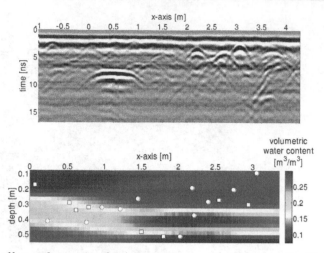

Fig. 9. Constant offset radar section (top) over an area with natural and artificial diffracting objects (metal bars, stones, roots) and the deduced water distribution based on the fit of diffraction hyperbolas (bottom). The circles and squares indicate the positions of objects that caused the diffractions used for the analysis (with permission from Schmalholz, 2007).

There are two principle modes by which a ground wave measurement can be carried out. The first is to perform a moveout (MO, WARR) or a CMP measurement (Overmeeren et al., 1997; Huisman et al., 2001). When plotting the traveltime versus the distance to the transmitter and receiver antennas, the airwave and the ground wave onsets form a straight line whose slope corresponds to the inverse of the wave velocity in air $1/c_0$ and soil $1/v_{soil}$, respectively (Fig. 10, left part, $x < x_{opt}$). The extension of the air and ground wave (dashed lines in Fig. 10) intersect at the origin. The second method is to carry out a constant offset (CO) measurement by measuring lateral changes in the velocity of the ground wave (Grote et al., 2003) (Fig. 10, right part, $x > x_{opt}$). In this mode, the water content distribution along a profile can be deduced rapidly. A combination of both methods was proposed by Du (1996) and is the most appropriate method to date.

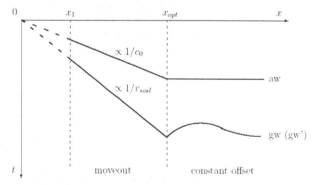

Fig. 10. Schematic traveltime diagram of a ground wave measurement consisting of moveout measurements from $x_1$ to $x_{opt}$, followed by a constant offset measurement for $x > x_{opt}$ (aw: air wave, gw: ground wave).

The ground wave technique is suited to map the water content of the topsoil over wide areas and to deduce its lateral distribution. The exact depth of investigation of the ground wave is still an object of research and is a function of antenna frequency, antenna separation and soil permittivity (e.g., Du, 1996; Galagedara et al., 2005). Fig. 11 shows an example of the result of the technique applied to a location with sandy soil that had been used as grassland. The area from $x = 9$ to 11 m was irrigated and shows higher water contents. TDR measurements were carried out every 20 cm along the same profile and show similar results.

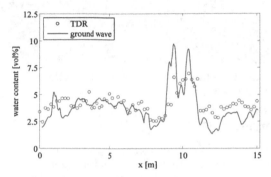

Fig. 11. Water contents obtained from ground wave field measurement of a sandy soil and results of TDR measurements. The area from 9 to 11 m has been irrigated and shows higher water contents.

### 3.6 Reflection coefficient at the soil surface

The reflection coefficient $R$ of an incident wave normal to an interface is described in Eq. 18. Regarding the interface air-soil with $\varepsilon_{air} = 1$, we can deduce the coefficient of reflection $R$ and, thus, the permittivity and the water content of the upper centimetres of the soil:

$$\varepsilon_{soil} = \left( \frac{1 - R}{1 + R} \right)^2$$

(28)

This principle is also used to determine soil moisture in active remote sensing techniques over large areas (e.g., Ulaby et al., 1986). Fig. 12 shows a layout with an air-launched horn antenna mounted on a sledge so that it can be moved along profiles with a fixed distance from the soil. An example of this technique is demonstrated in Section 4.1.

### 3.7 Further techniques

In addition to these commonly used techniques, there are a variety of other techniques that can be used to deduce soil-water content by means of GPR. For example, Bradford (2007) analysed the frequency dependence of wave attenuation in the high-frequency range to assess water content and porosity from GPR measurements. Igel et al. (2001) and Müller et al. (2003) used guided waves that travel along a metallic rod that is lowered into a borehole. Van der Kruk (2010) analysed the dispersion of waves that propagate in a layered soil under certain conditions and Minet et al. (2011) used a full-waveform inversion of off-ground horn antenna data to map the moisture distribution of a horizontally layered soil.

Fig. 12. Experimental setup to determine the coefficient of reflection. A 1-GHz horn antenna is mounted on a sledge 0.5 m above the ground.

## 4. Application examples of GPR

In this section, two examples of GPR measurements and analyses applied for monitoring irrigation and subsequent events are given. Both examples demonstrate the ability of GPR to measure and monitor changes in the soil water content. The first example analyses reflections at the soil surface (see Section 3.6) and the second is transmission measurements between boreholes (see Section 3.1).

### 4.1 Monitoring of soil water content variation by GPR during infiltration

GPR measurements were carried out after irrigation to see its capability in monitoring the changes in soil dielectric properties and their spatial variation. This work demonstrates that the obtained GPR data can be used to describe the soil water distribution.

### 4.1.1 Experimental setup

An irrigation experiment was carried out on an outdoor test site in Hannover, Germany. The texture of the soil at the site is medium sand and the soil has a high humus content of 6.3%. The total pore volume (i.e., the maximum water capacity shortly after heavy rainfall or irrigation) of the soil is estimated to be 55 vol%. The field capacity (i.e., the maximum water content that can be stored against gravity) is approximately 21 vol%, which corresponds to a relative dielectric permittivity of 10.7 according to Topp's equation (Topp et al., 1980). The ground surface was relatively flat.

A one-square-metre area was irrigated at a rate of approximately 12 litres per minute for approximately one hour, i.e., 720 litres of water in total. An excess of water was applied to ensure that the pore system was filled to its maximum water capacity. The relative permittivity of the soil before and immediately after the irrigation was 4 and 20.8, which corresponds to a water content of 5.5 vol% and 35.5 vol%, respectively. Because the test was carried out during the summer and there had been no rainfall for more than a half month prior to the test, the soil was assumed to be in its driest natural condition prior to the irrigation.

A frequency-domain radar system was employed and operated at a frequency range of 0.5-4.0 GHz. Antennas with a constant separation of 6 cm were fixed at a height of 5 cm above the ground surface and scanned in 1D every two minutes on an approximately 1-m-long profile with the help of a scanner. Additionally, the relative permittivity was monitored at one location off the GPR scanning line by TDR every half minute. The TDR probes are 10 cm long and were stuck vertically in the topsoil. The configuration of the measurements is illustrated in Fig. 13. The data collection was continued after the irrigation was stopped. The relative permittivity measured by TDR and soil water content at the end of the irrigation were 14 and 26 vol%, respectively (see Fig. 16).

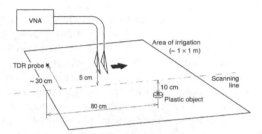

Fig. 13. Schematic illustration of the experimental setup.

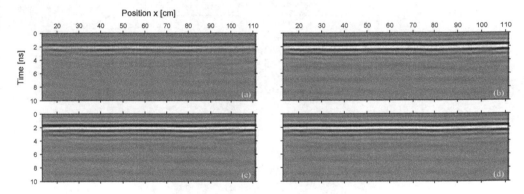

Fig. 14. Radar profiles acquired (a) before irrigation, (b) when irrigation was stopped, (c) one hour after irrigation stopped, and (d) two hours after irrigation stopped.

### 4.1.2 Data analysis and estimation of soil conditions

Some of the radar profiles obtained in the experiment are shown in Fig. 14. These profiles are plotted with the same colour scale. The reflections from the ground surface at 2 ns are weaker before irrigation (Fig. 14a) than after (Fig. 14b-d). This is because the reflection coefficient of the ground surface was increased by irrigation and the surface reflection became stronger. Thus, by analysing the surface reflection, it is possible to retrieve the soil water content at the soil surface as described in Section 3.5.

Fig. 15 shows waveforms acquired at the same location but at different times. The surface reflections appear at around 2 ns, but the peak amplitudes $E^r$ are different. The amplitude of the reflection is expressed with a reflection coefficient $R$, following Eq. 14. The reflection coefficient of an obliquely incident electromagnetic wave with an electric field parallel to the interface is given by

$$R = \frac{Z_2 \cos\theta_1 - Z_1 \cos\theta_2}{Z_2 \cos\theta_1 + Z_1 \cos\theta_2}$$

(29)

where $\theta_1$ and $\theta_2$ are the angles of the incident and transmitted waves, respectively. The incident angle is determined by the measurement geometry. $Z_1$ and $Z_2$ are the intrinsic impedances of the first medium (air) and second medium (soil), respectively. The intrinsic impedance of air and incident angle are known, and the reflection amplitude was measured. Therefore, the intrinsic impedance of soil $Z_2$, the dielectric permittivity and the water content can be calculated from the above relationships. The only unknown is the strength of the incident wave $E^i$, which is often measured with a metal plate with $R = 1$ as calibration (Serbin & Or, 2003; Igel, 2007). In this experiment, the incident wave strength was obtained using a puddle on the ground caused by the excess water supply during the irrigation. Assuming the permittivity of the water is 81, the incident wave strength can also be obtained with the above equations. If there are changes in the surface topography, the amplitude of the incident wave changes at different locations because of the antenna radiation pattern, i.e., changes in spreading loss. However, the ground surface in this experiment was almost flat, as can be seen from the traveltimes of the reflected wave in Fig. 14, which are almost constant along the profile, so the effects of topography were neglected.

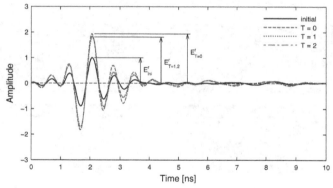

Fig. 15. Waveforms sampled before irrigation (in the dry condition, labelled "initial"), when irrigation was stopped (T = 0), and one and two hours after irrigation was stopped.

The estimated dielectric permittivity from GPR data at three positions along the profile are shown in Fig. 16 as a function of time after irrigation in comparison to the TDR measurements. The corresponding volumetric water content is also provided on the right axis using Topp's equation (Topp et al., 1980). Before irrigation, the permittivity values estimated from GPR measurements and measured by TDR are similar, but the estimated ones are slightly lower. This is because the TDR measures permittivity averaged over the measurement volume, which is about 10 cm deep, while the estimate using GPR is for a shallower region, perhaps a centimetre deep. In its dry condition, the ground surface may be drier than deeper regions, and the method could retrieve this difference. Furthermore, a lower bulk density in a shallower region leads to a lower bulk permittivity. At the time that irrigation was stopped, the water content measured by TDR was about 35 vol%, while the water content estimated by GPR was higher than 50 vol%. This indicates that within very shallow soil, most of the pore space was filled with water. After the irrigation was stopped, the measured water content slowly decreased exponentially to about 25 vol%, which is consistent with our estimate of the field capacity. The estimated water content also decreased with time, but the decrease was much more rapid than for the water content measured with TDR. This is also caused by the different sampling depths of the two methods: the shallower layer mapped by GPR loses water faster than the deeper soil horizons measured with TDR.

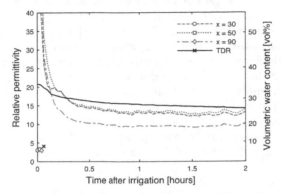

Fig. 16. The dielectric permittivity of the uppermost 10 cm of the soil measured by TDR (solid line) and permittivity of the ground surface estimated from GPR data, plotted as a function of time after the irrigation was stopped. Plots at the zero-time axis indicate permittivity before irrigation, i.e., the dry condition. The right axis gives the corresponding volumetric water content using Topp's equation (Eq. 22, Topp et al., 1980).

The dielectric permittivity of the ground surface can also be estimated depending on the position along the profile. The left side of Fig. 17 shows the spatial variation in permittivity and water content at different times as a function of the position. The water content, as well as its spatial variation, were not high in the dry condition, but both were increased by irrigation. The spatial distributions before and after irrigation are similar but not exactly the same. As shown in Fig. 16, most of the decrease in water content in the shallow region occurred within the first 30 minutes, and the calculated spatial water distribution at one and two hours after irrigation ceased are almost the same.

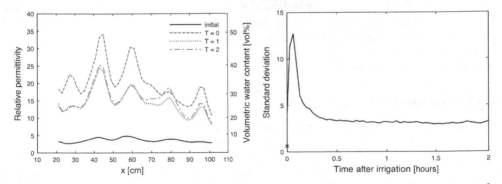

Fig. 17. Left: the spatial variation in dielectric permittivity of the ground surface estimated from GPR data. Right: the standard deviation of the estimated relative permittivity in space. The cross at time zero indicates the value before irrigation.

By further analysing the spatial variation of dielectric permittivity and water content, the soil water distribution can be understood. The right side of Fig. 17 shows the standard deviation of permittivity plotted as a function of time. The variation was very small for the dry condition. By adding water, the variation increased and continued to increase for about 10 minutes after irrigation was stopped. It then decreased and became stable but was still higher than for the dry condition. The result can be interpreted as water that had infiltrated and percolated into the ground during and shortly after irrigation, but the percolation velocity varies spatially on a small scale, depending on differences in bulk density, texture, humus content and related pore volume, i.e., water was drained faster in areas with higher proportions of transmission pores. The variation in percolation velocity caused a variation in water content and an increase in the standard deviation of permittivity for a short period of time after irrigation was stopped. Areas with a smaller pore volume may have stored more water than their field capacity for a short time, but the excess water subsequently percolated out. This behaviour led to a decrease in variation in permittivity. After some time, all sections stored as much water as the field capacity and the variation became stable, but remained higher than at the dry condition.

## 4.2 Infiltration monitoring by borehole GPR

An artificial groundwater recharge test was carried out in Nagaoka City, Japan (Kuroda et al., 2009). Time-lapse cross-hole measurements were performed at the same time to monitor the infiltration process in the vadose zone.

### 4.2.1 Field experiment

The borehole GPR measurement carried out during the infiltration experiment is schematically illustrated in Fig. 18. The top 2 m of soil consisted of loam, and the subsoil was sand and gravel. The groundwater table was located approximately 10 m below the ground surface. During the experiment, water was injected from a 2 x 2 m tank with its base set at a depth of 2.3 m. The total volume of water injected into the soil was 2 m³, requiring approximately 40 minutes for all water to flow from the tank. Two boreholes were located on both sides of the tank with a separation of 3.58 m. Using these boreholes, GPR data were acquired in a ZOP configuration. In this mode, both the transmitter and receiver antennas were lowered to a common depth. Data were collected every 0.1 m at depths of 2.3-5.0 m. This required about 2 minutes to cover the whole depth range. A total of 25 profiles were obtained during the experiment, which lasted 322 minutes.

### 4.2.2 Estimation of percolation velocity

Three radar profiles acquired from this experiment are shown in Fig. 19. The first arrival times range from 36-38 ns in the initial, unsaturated state (Fig. 19a), and 42-44 ns in the final state (Fig. 19c), which is considered to be fully saturated. In the intermediate state of 51-53 minutes after infiltration, the first arrival times are almost identical to the initial state at depths below 4 m (Fig. 19b). Shallower than 4 m depth, the first arrival times are delayed: the shallower the depth, the greater the traveltime. This delay is caused by increased wetting in the vadose zone (transition zone).

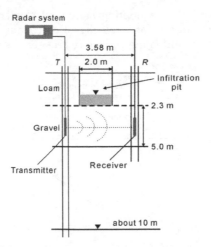

Fig. 18. Schematic sketch of the artificial ground water recharge test in the vadose zone.

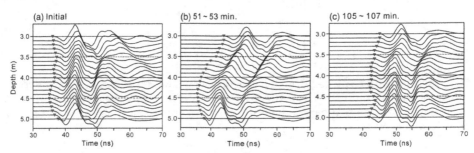

Fig. 19. Radar profiles observed during the infiltration test: (a) before infiltration, (b) 52 min after infiltration, and (c) 106 min after infiltration. Triangles show the picked first arrival times.

The infiltration process may not be 1D because the vadose zone is heterogeneous. Since the standard ZOP method relies on determining the velocity of an electromagnetic wave that follows a straight path from the transmitter to the receiver, the infiltration process is assumed to be 1D. Fig. 20 shows vertical profiles of the first arrival times. In these profiles, the volumetric water content at a specific depth is estimated based on the following procedure. By assuming a straight raypath, a first arrival time is used to calculate the velocity $v$ using $v = d/t$, where $d$ is the offset distance between the transmitter and the receiver, and $t$ is the traveltime. By assuming that frequency-dependent dielectric loss is small (Davis and Annan, 1989), the apparent dielectric constant $\varepsilon$ is obtained using Eq. 2. Finally, the volumetric water content $\theta$ can be estimated by substituting $\varepsilon$ into the empirical Topp's equation (Eq. 22, Topp et al., 1980).

The spatial and temporal variations of volumetric water content are derived using all first arrival times collected during the infiltration experiment (Fig. 21). This illustration clearly shows the movement of the wetting front in the test zone during the infiltration process. The water content varies sharply from the unsaturated state (lower left side) to the saturated state (upper right side). The zone exhibiting significant changes can be interpreted as a transition zone. The average downward velocity of percolating water in the test zone is estimated to be about 2.7 m/h and is the boundary between the unsaturated and transition zones (Fig. 21).

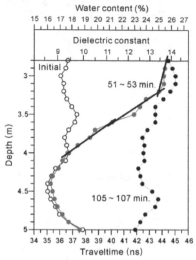

Fig. 20. First arrival times picked from the time-lapse ZOP radar profiles at three different conditions of the infiltration experiment. EM-wave velocities estimated using first arrival times are transformed into apparent dielectric constants and further converted into volumetric water contents.

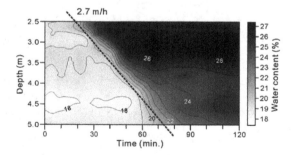

Fig. 21. A map showing the spatiotemporal variations in volumetric water content in the test zone during the infiltration process. The infiltrated water penetrated vertically with a velocity of about 2.7 m/h.

## 5. Conclusion

In this chapter, the basics of GPR and the principles for measuring soil water content are discussed. Two examples are given where GPR was used to monitor the temporal changes in soil water content following irrigation. The first example describes analysis of the reflection amplitude of electromagnetic waves, which depends on the dielectric permittivity and the water content of soil. This technique is very fast and can be used to obtain results with a very high spatial and temporal resolution. The second example utilised the fact that the velocity of the electromagnetic waves is determined by the dielectric permittivity and the water content of soil. Although this method requires boreholes to carry out a transillumination measurement, it can capture vertical changes in soil water content. These examples clearly demonstrated the capability of GPR to measure water-related soil properties. Compared to other methods for the measurement of soil water content, GPR can measure a larger area easily and quickly with a high spatial and temporal resolution. As shown in Section 3, there are a variety of techniques to measure or estimate the soil water content. Most of them are completely non-invasive, which is a very important feature for monitoring soil water content at the same location. One can chose the most suitable method according to the possible measurement setup and the desired type of result. Therefore, GPR has a great potential for use in investigations of irrigation and soil science.

## 6. List of symbols

$c$        Velocity of light in free space          m/s
$d$        Depth    m
$f$        Frequency        Hz
$k$        Wavenumber      1/m
$t$        Time      s
$\tan\delta$    Loss tangent
$v$        Propagation velocity of electromagnetic waves          m/s
$E$        Electric field strength of an electromagnetic wave      V/m
$E_0$      Original electric field strength of an electromagnetic field        V/m
$E^i$      Incident electric field strength        V/m
$E^r$      Reflected electric field strength        V/m
$E^t$      Transmitted electric field strength   V/m
$R$        Reflection coefficient
$S_w$      Water saturation
$T$        Transmission coefficient
$Z$        Intrinsic impedance        $\Omega$
$\alpha$   Attenuation constant        Np/m
$\alpha'$  Attenuation constant        dB/m
$\beta$    Phase constant    1/m
$\delta$   Skin depth        m
$\varepsilon$     Absolute dielectric permittivity      F/m
$\varepsilon'$    Real part of dielectric permittivity  F/m
$\varepsilon''$   Imaginary part of dielectric permittivity        F/m
$\varepsilon_0$   Absolute dielectric permittivity of free space F/m
$\varepsilon_e$   Complex effective permittivity        F/m
$\varepsilon_r$   Relative dielectric permittivity
$\varepsilon_s$   Low frequency static permittivity    F/m
$\varepsilon_\infty$   High frequency permittivity        F/m
$\phi$     Porosity
$\lambda$  Wavelength        m

| | | |
|---|---|---|
| $\mu$ | Absolute magnetic permeability | H/m |
| $\mu_0$ | Absolute magnetic permeability of free space | H/m |
| $\mu_r$ | Relative magnetic permittivity | |
| $\theta_v$ | Volumetric water content | |
| $\sigma$ | Electric conductivity | S/m |
| $\sigma'$ | Real part of electric conductivity | S/m |
| $\sigma''$ | Imaginary part of electric permittivity | S/m |
| $\sigma_q$ | Interface conductivity | S/m |
| $\tau$ | Relaxation time | s |
| $\omega$ | Angular frequency | rad/s |

# 7. References

Annan, A.P. (2005). Ground penetrating radar in near-surface geophysics, In: Near-Surface Geophysics, Investigations in Geophysics, No. 13, Society of Exploration Geophysics, Butler, D.K., pp.357-438, ISBN 1-56080-130-1, Tulsa, OK

Annan, A.P. (2009). Electromagnetic principles of ground penetrating radar, In: Ground Penetrating Radar: Theory and Applications, Jol, H.M., pp. 3-40, Elsevier, ISBN 978-0-444-53348-7, Amsterdam, The Netherlands

Archie, G.E. (1942). The electrical resistivity log as an aid in determining some reservoir characteristics. Petroleum Transactions of AIME, Vol. 146, pp. 54-62

Binley, A., Winship, P., Middleton, R., Pokar, M. & West, J. (2001). High-resolution characterization of vadose zone dynamics using cross-borehole radar. Water Resources Research, Vol. 37, pp. 2639-2652

Bradford, J.H. (2007). Frequency-dependent attenuation analysis of ground-penetrating radar data. Geophysics, Vol. 72, pp. J7-J16

Bradford, J.H. (2008). Measuring water content heterogeneity using multifold GPR with reflection tomography. Vadose Zone Journal, Vol. 7, pp. 184-193

Cassidy, N.J. (2009). Electrical and magnetic properties of rocks, soils, and fluids, In: Ground Penetrating Radar: Theory and Applications, Jol, H.M., pp. 41-72, Elsevier, ISBN 978-0-444-53348-7, Amsterdam, The Netherlands

Daniels, D.J. (Ed.). (2004). Ground Penetrating Radar (2nd Edition), IEE, ISBN 0-86341-360-9, London

Davis, J.L. & Annan, A.P. (1989). Ground-penetrating radar for high resolution mapping of soil and rock stratigraphy. Geophysical Prospecting, Vol. 37, pp. 531-551

Dix, C.H. (1955). Seismic velocities from surface measurements. Geophysics, Vol. 20, pp. 68-86

Du, S. (1996). Determination of Water Content in the Subsurface with the Ground Wave of Ground Penetrating Radar. Dissertation, Ludwig-Maximilians-Universität, München, Germany

Galagedara, L.W., Redman, J.D., Parkin, G.W., Annan, A.P. & Endres, A.L. (2005). Numerical modeling of GPR to determine the direct ground wave sampling depth. Vadose Zone Journal, Vol. 4, pp. 1096-1106

Greaves, R.J., Lesmes, D.P., Lee, J.M. & Toksoz, M.N. (1996). Velocity variations and water content estimated from multi-offset ground penetrating radar, Geophysics, Vol. 61, pp. 683-695

Grote, K., Hubbard, S. & Rubin, Y. (2003). Field-scale estimation of volumetric water content using ground-penetrating radar ground wave techniques. Water Resources Research, Vol. 39, pp. 1321-1333

Huisman, J.A., Sperl, C., Bouten, W. & Verstaten, J.M. (2001). Soil water content measurements at different scales: accuracy of time domain reflectometry and ground-penetrating radar. Journal of Hydrology, Vol. 245, pp. 48-58

Huisman, J.A., Hubbard, S.S., Redman, J.D. & Annan, A.P. (2003). Measuring soil water content with ground penetrating radar: a review. Vadose Zone Journal, Vol. 2, pp. 476-491

Igel, J., Schmalholz, J., Anschütz, H.R., Wilhelm, H., Breh, W., Hötzl, H. & Hübner C. (2001). Methods for determining soil moisture with the ground penetrating radar (GPR). Proceedings of the Fourth International Conference on Electromagnetic Wave Interaction with Water and Moist Substances, Weimar, Germany.

Igel, J. (2007). On the Small-Scale Variability of Electrical Soil Properties and Its Influence on Geophysical Measurements. PhD Thesis, Johann Wolfgang Goethe-Universität, Frankfurt am Main, Germany

Kaatze, U. (1989). Complex permittivity of water as a function of frequency and temperature. Journal of Chemical and Engineering Data, Vol. 34, pp. 371-374

Knight, R.J. & Endres, A.L. (2005). An introduction to rock physics principles for near-surface geophysics, In: Near-Surface Geophysics, Investigations in Geophysics, No. 13, Society of Exploration Geophysics, Butler, D.K., pp.357-438, ISBN 1-56080-130-1, Tulsa, OK

Knödel, K., Lange, G., & Voigt, H.-J. (Eds.). (2007). Environmental Geology, Handbook of Field Methods and Case Studies, Springer, ISBN 978-3-540-74669-0, Berlin

Kuroda, S., Jang, H. & Kim, H.J. (2009). Time-lapse borehole radar monitoring of an infiltration experiment in the vadose zone. Journal of Applied Geophysics, Vol. 67, pp. 361-366

Lampe, B. & Holliger, K. (2003). Effects of fractal fluctuations in topographic relief, permittivity and conductivity on ground-penetrating radar antenna radiation. Geophysics, Vol. 68, pp. 1934-1944

Minet, J., Wahyudi, A., Bogaert, P., Vanclooster, M., & Lambot, S. (2011). Mapping shallow soil moisture profiles at the field scale using full-waveform. Geoderma, Vol. 161, pp. 225-237

Müller, M., Mohnke, O., Schmalholz, J. & Yaramanci, U. (2003). Moisture assessment with small-scale geophysics - the Interurban project. Near Surface Geophysics, Vol. 1, pp. 173-181

Noborio, K. (2001). Measurement of soil water content and electrical conductivity by time domainreflectometry: a review. Computers and electronics in agriculture, Vol. 31, 213-237

Overmeeren, R.A., Sariowan, S.V. & Gehrels, J.C. (1997). Ground penetrating radar for determining volumetric soil water content; results of comparative measurements at two test sites. Journal of Hydrology, Vol. 197, pp. 316-338

Preetz, H., Altfelder, S. & Igel, J. (2008). Tropical soils and landmine detection – an approach for a classification system. Soil Science Society of America Journal, Vol. 72, pp. 151-159

Robinson, D.A., Jones, S.B., Wraith, J.M., Or, D. & Friedmann, S.P. (2003). A review of advances in dielectric and electrical conductivity measurement in soils using time domain reflectometry. Vadose Zone Journal, Vol. 2, pp. 444-475

Schmalholz, J., Stoffregen, H., Kemna, A. & Yaramanci, U. (2004). Imaging of Water Content Distributions inside a Lysimeter using GPR Tomography. Vadose Zone Journal, Vol. 3, pp. 1106-1115

Schmalholz, J. (2007). Georadar for small-scale high-resolution dielectric property and water content determination of soils. Dissertation, Technical University of Berlin, Germany

Serbin, G. & Or, D. (2003). Near-surface soil water content measurements using horn antenna radar: methodology and overview. Vadose Zone Journal, Vol. 2, pp. 500-510

Shen, L.C., Savre, W.C., Price, J.M. & Athavale, K. (1985). Dielectric properties of reservoir rocks at ultra-high frequencies, Geophysics, Vol. 50, pp. 692-704

Stoffregen, H., Zenker, T., Wessolek, G. (2002). Accuracy of soil water content measurements using ground penetrating radar: Comparison of ground penetrating radar and lysimeter data. Journal of Hydrology, Vol. 267, pp. 201-206

Takahashi, K., Preetz, H. & Igel, J. (2011). Soil properties and performance of landmine detection by metal detector and ground-penetrating radar – Soil characterisation and its verification by a field test. Journal of Applied Geophysics, Vol. 73, pp. 368-377

Topp, G.C., Davis, J.L. & Annan, A.P. (1980). Electromagnetic determination of soil water content: measurements in coaxial transmission lines. Water Resource Research, Vol. 16, pp.574-582

Tronicke, J., Tweeton, D.R., Dietrich, P. & Appel, E. (2001). Improved crosshole radar tomography by using direct and reflected arrival times. Journal of Applied Geophysics, Vol. 47, pp. 97-105

Ulaby, F.T., Moore, R.K. & Adrian, K.F. (1986). Microwave Remote Sensing: Active and Passive, volume 3: From Theory to Applications. Artech House, ISBN 0-89006-192-0, Norwood, MA

van der Kruk, J., Jacob, R.W. & Vereecken, H. (2010). Properties of precipitation-induced multilayer surface waveguides derived from inversion of dispersive TE and TM GPR data. Geophysics, Vol. 75, pp. WA263-WA273

Yilmaz, O. (2000). Seismic data analysis - processing, inversion, and interpretation of seismic data (2nd edition). Society of Exploration Geophysicists, ISBN 978-0931830464, Tulsa, OK

# 9

# Using Wireless Sensor Networks for Precision Irrigation Scheduling

John D. Lea-Cox
*Professor, Department of Plant Science and Landscape Architecture*
*University of Maryland, College Park*
*USA*

## 1. Introduction

Worldwide, irrigation uses about 69% of available freshwater resources (Fry, 2005). In the United States, 82% of freshwater resources are used for irrigation purposes. Major concerns on future planetary freshwater resources are the effects of climate change on changing sea temperature and levels, annual snowpack, drought and flood events, as well as changes in water quality, and general ecosystem vulnerabilities (US Global Change Research Program, 2011). Changes in the extreme climatic events are more likely to occur at the regional level than show in national or global statistics. The unpredictability of climatic events is of key concern to farmers in all countries, since the availability and cost of irrigation water is likely to be compounded by increased regulations and competition. Over the past 50 years, the urban demand for freshwater in the United States has also been increasing (Hutson et al. 2004), while the quality of both surface and groundwater has been decreasing due to pollution from both point and nonpoint sources (Secchi et al. 2007). Nitrogen, phosphorus and many other inorganic and organic pollutants such as pesticides and herbicides are being found at increasing concentrations in groundwater under agricultural areas (Guimerà, 1998). As demands on water and the cost of purification increase, the cost of freshwater resources will increase and the availability will likely decrease for agriculture. Population growth in the 20th century increased by a factor of three while water withdrawals increased by a factor of seven during the same time, with little hope of these rates slowing in the near future (Agarwal et al. 2000).

In view of increased competition for resources and the need for increased agricultural production to ensure national and global food security, it is clear that we need to increase our efficiency of irrigation water use, to adapt to these changing conditions. Not only do we need to increase the overall efficiency of irrigation water use to optimize crop yields, but there is also a need to provide farmers with better information on root zone water availability and daily crop water use, especially at critical times during flowering, fruit set and fruit or seed development. Although crop yield is oftentimes related to water use, most growers don't know the water requirement of the crop they grow at any real level of precision. Since irrigation costs in developed countries are usually a small fraction of total production costs, there are few incentives for growers to optimize their use of irrigation water. Therefore, the amount of water applied is mostly based on the availability, rather than actual crop water needs (Balendock et al., 2009). The development of precision (low volume) irrigation systems  has played a major role in reducing the water required to maintain yields for high-value crops, but this has also highlighted the need for new methods for accurate irrigation scheduling and control (Jones, 2008). For high-value horticultural crops, there is also significant interest in using precision irrigation as a tool to increase harvest quality through regulated deficit irrigation, and to reduce nutrient loss and fungal disease pressures. In the near future, farmers will likely have to make decisions on how to optimize water use with crop yield, in order to remain competitive. Achievement of any optimal irrigation capability will depend not only on the use of precision irrigation systems, but also on the tools that can help the farmer monitor and automate irrigation scheduling, applying water precisely to satisfy crop water requirements.

## 1.1 Scope of the chapter

The intent of this chapter is not to provide the reader with an exhaustive review of the sensor network development literature. A simple online search of the keywords "wireless, sensor, network, irrigation" provides links to over 1500 journal articles just within the engineering and biological fields. However, two recent articles do provide excellent current reviews (Ruiz-Garcia et al., 2009) and practical advice (Barrenetxea et al., 2008) for readers wanting more explicit engineering and technical advice on the development and deployment of wireless sensor networks for irrigation, environmental and other (animal and food safety) applications. It is apparent from these and many other articles that the development of operational wireless sensor networks (WSNs) for large-scale outdoor deployment has many challenges; an excellent discussion of many of the pitfalls is given by Barrenetxea et al., (2008). For many of the reasons they and others have outlined, most research in the field of sensor networks for irrigation scheduling has focused on the technical challenge of gathering reliable data from wide area networks. Barrenetxea et al. (2008) note that there are two primary components for successful WSN deployments: (1) gathering the data and (2) exploiting the data. Generally, the engineering component of any WSN project is tasked with providing hardware that reliably accomplishes the first task. However, understanding the biological and/or environmental domain is vital to maximize the trustworthiness of sensor data. Interdisciplinary projects and partnerships are more likely to have a greater chance of success, since the primary objective of all WSN's is to gather data for a specific use, and all partners are focused on that task. However, if we are going to successfully commercialize and deploy sensor networks on farms, the *end-user* must be involved during all stages of the project: from node deployment, to sensor placement and calibration, through to data analysis and interpretation (Tolle et al, 2005; Lea-Cox et al., 2010a). It is this part of the process that has often received scant attention from researchers and developers; the successful integration of sensor networks and decision support systems (software tools) is probably one of the greatest barriers to successful implementation and adoption of these systems by farmers. For this reason, any tools that are developed should be thoroughly vetted by the end-user for ease-of-use, interpretation and applicability. Perhaps most importantly, we should learn from past mistakes where various water-saving technologies have often not achieved any real economic benefit for the grower, in terms of water savings, improved yield, labor cost or other use. Sometimes technology merely adds another "management layer" that requires additional expertise to interpret and use the information, in order to make a decision. We therefore need to bear these considerations in mind when we develop any irrigation scheduling system that aims to improve upon current irrigation management techniques.

This chapter will firstly summarize the pros and cons of certain sensors and techniques that provide promise for use in WSNs for precision irrigation. It will review WSN deployment and progress, but focus primarily on intensive nursery and greenhouse production, since these environments provide some extreme challenges with spatial and temporal sensor measurement, to accurately predict plant water use. We recognize that there are many aspects of plant physiology that provide both feedback and feedforward mechanisms in regulating plant water use, and these may radically change in a crop, pre- and post-anthesis. This chapter is not focused on these challenges; it will however attempt to illustrate the potential of sensor networks to provide real-time information to both farmers and researchers — often at a level of precision that provides keen insights into these processes. Our research and development team (Lea-Cox et al, 2010b) is actively working on deploying and integrating WSNs on farms, but concurrently developing the advanced hardware and software tools that we need for precision irrigation scheduling in intensive horticultural operations. We will illustrate some of our progress with these WSN's in container-production and greenhouse environments, as well as in field (soil-based) tree farms which have soil water dynamics more akin to field orchard environments. Finally, we will discuss challenges and opportunities that need to be addressed to enable the widespread adoption of WSN's for precision irrigation scheduling.

## 1.2 Intensive production system irrigation scheduling

The most widespread use of automated irrigation scheduling systems are in intensive horticultural, and especially in greenhouse or protected environments (Jones, 2008). Currently, many greenhouse and nursery growers base their irrigation scheduling decisions on intuition or experience (Bacci et al., 2008; Jones, 2008; Lea-Cox et al., 2009). Oftentimes the most basic decision is — "do I need to irrigate today?" While this question could seem trivial, plant water requirements vary by species, season and microclimate, and depend upon any number of environmental and plant developmental factors that need to be integrated on a day-to-day basis. Add to these factors the number of species grown in a 'typical' nursery or greenhouse operation (oftentimes >250 species; Majsztrik, 2011), the variety of container sizes (i.e. rooting volume, water-holding capacity) and the length of crop cycles (a few weeks to several years), it quickly becomes obvious why precision irrigation scheduling in these types of operations is extremely difficult (Lea-Cox et al., 2001; Ross et al, 2001). If done well, daily irrigation decisions take a lot of time and an irrigation manager often faces complex decisions about scheduling, which requires integrating knowledge from many sources. Although these intuitive methods for irrigation scheduling can give good results with experience, they tend to be very subjective with different operators making very different decisions. Many times, even experienced managers make an incorrect decision, i.e., they irrigate when water is not required by the plant. It is also surprising how many "advanced" irrigation scheduling systems automate irrigation cycles *only* on the basis of time, without any feedback-based sensor systems. Thus, even with advanced time-based systems, the decision to irrigate is again based solely on the operator's judgment and the time taken to evaluate crop water use and integrate other information, e.g. weather conditions during the past few days and immediate future.

There are many sensor technologies that have been used over the years to aid this decision process. Various soil moisture measurement devices are available, e.g. tensiometers, gypsum blocks and meters that directly sense soil moisture; additionally pan evaporation, weather station or satellite forecast data can be incorporated into evapotranspiration ($E_T$) models. However, the widespread adoption of most of this technology has not occurred in the nursery and greenhouse industries, for good reasons. Many sensing technologies which were originally engineered for soil-based measurements have been applied to soilless substrates. Many have failed, largely because these sensors did not perform well in highly porous substrates, since porosity is an important physical property that is necessary for good root growth in containers (Bunt, 1961). Even when a technology has been adapted successfully to container culture (e.g. low-tension tensiometers), often the technology has been too expensive for wide-scale adoption, difficult to incorporate into WSNs, or there have been precision or maintenance issues. Cost and ease of use are key aspects to the adoption and use of any tool by growers, who are often time-limited.

## 1.3 Wireless sensor network development objectives

It was imperative to establish a list of global objectives for the development of WSN tools and strategies for our project (Lea-Cox et al, 2010b). Jones (2008) documented the features of an 'ideal' irrigation scheduling system for intensive horticultural production systems. He noted that any system should be (1) sensitive to small changes, whether in terms of soil moisture content, evaporative demand, or plant response; (2) respond rapidly to these changes, allowing for continual monitoring and maintenance of optimal water status and responding in "real time" to changing weather conditions; (3) readily adaptable to different crops, growth stages or different horticultural environments without the need for extensive recalibration; (4) robust and reliable; (5) user-friendly, requiring little user training; (6) capable of automation, thus reducing labor requirements, and (7) low cost, both in terms of purchase and running costs.

In addition to these universal requirements, Lea-Cox et al. (2008) proposed a number of more specific WSN requirements, where (1) users should be able to rapidly deploy sensors in any production area, to maximize utility and minimize cost; (2) sensor networks should be scalable, thereby allowing an operation to begin with a small, low-cost system and expand/improve the network over time; (3) nodes (motes) should have low power (battery) requirements, preferably with rechargeable power options; (4) sensor data should be reliably transmitted using wireless connections to the base station computer (or internet) with little or no interference over at least 1000m; (5) the software interface should automatically log and display real-time data from the sensor nodes, in a form that provides the user with an easily interpreted summary of that data, preferably as a customizable graphical output (6) any software control functions should include relatively sophisticated decision tools and discretionary options, to allow for maximum flexibility in scheduling / actuating irrigation solenoids or other control devices.

These engineering objectives are the foundation for our specific scientific, engineering and socio-economic objectives (Lea-Cox, 2010b), to: (1) further develop and adapt commercially-available wireless sensor network hardware and software, to meet the monitoring and control requirements for field (soil-based), container (soilless) production and environmental (green roof) systems; (2) determine the performance and utility of soil moisture and electrical conductivity sensors for precision irrigation and nutrient management; (3) determine spatial and temporal variability of sensors, to minimize the numbers of sensors required for different environments at various scales; (4) integrate various environmental sensors into WSNs to enable real-time modeling of microclimatic plant $E_T$; (5) integrate soil and environmental data into species-specific models to better predict plant and system water use; (6) develop best management practices for the use of sensors, working with commercial growers to capture needs-based issues during on-farm system development; (7) quantify improvements in water and nutrient management and runoff, plant quality, and yield; (8) evaluate the private and public economic and environmental impacts of precision sensor-controlled practices; (9) identify barriers to adoption and implementation of these practices; and (10) engage growers and the industry on the operation, benefits and current limitations of this sensor / modeling approach to irrigation scheduling and management.

## 2. Irrigation sensing approaches

The main approaches to irrigation scheduling in soils and the techniques available have been the subject of many reviews over the years. Specific reviews have concentrated on measuring soil moisture (e.g. Dane & Topp, 2002; Bitelli, 2011), physiological measurements (e.g. Jones, 2004; Cifre et al., 2005) or water balance calculations (e.g. Allen et al., 1998). The conventional sensor-based approach has typically scheduled irrigation events on the basis of soil moisture status, whether using direct soil moisture measurements with capacitance or TDR-type sensors (Topp, 1985; Smith & Mullins, 2001), tensiometers (Smajstrla & Harrison, 1998) or soil-moisture water balance methods using daily $E_T$ estimates (Allen et al., 1998). Some automated greenhouse systems have used load cells for the estimation of daily plant water use (Raviv et al., 2000). However, these load cell systems have to be programmed to accurately correct for increasing total plant mass over the crop cycle, or to adjust to changes in wind and temperature changes, if deployed in outdoor environments. Nevertheless, if operated correctly, most of these systems enable much greater precision and improved water use efficiency over traditional time-based irrigation scheduling methods.

Jones (2004) summarized the main sensor techniques that are currently used for irrigation scheduling or which have the potential for development in the near future in some detail (Table 1). The current debate centers around using soil moisture sensing techniques, plant water sensing techniques or a combination of both techniques. Soil irrigation sensing approaches (Table 1) can either be based on direct measurement of soil moisture content (or water potential), or by using sensors to provide data for the water balance method, which accounts for inputs (rainfall, irrigation) and losses ($E_T$, run-off and drainage) from the system. The emphasis on using soil moisture content for irrigation decisions has been based on the perception that water availability in the soil is what limits plant transpiration, and that irrigation scheduling should replace the water lost by plant water uptake and evaporation from the rootzone (Jones, 2008).

| Measurement Technique | Advantages | Disadvantages |
| --- | --- | --- |
| I. Soil water measurement (a) Soil water potential (tensiometers, psychrometers, etc.)<br><br>(b) Soil water content (gravimetric; capacitance / TDR; neutron probe) | Easy to apply in practice; can be quite precise; at least water content measures indicate 'how much' water to apply; many commercial systems available; some sensors (especially capacitance and time domain sensors) readily automated | Soil heterogeneity requires many sensors (often expensive) or extensive monitoring program (e.g. neutron probe); selecting position that is representative of the root-zone is difficult; sensors do not generally measure water status at root surface (which depends on evaporative demand) |
| II. Soil water balance calculations (Require estimate of evaporation and rainfall) | Easy to apply in principle; indicate 'how much' water to apply | Not as accurate as direct measurement; need accurate local estimates of precipitation / runoff; evapotranspiration estimates require good estimates of crop coefficients (which depend on crop development, rooting depth, etc.); errors are cumulative, so regular recalibration needed |
| III. Plant 'stress' sensing (Includes both water status measurement and plant response measurement) | Measures the plant stress response directly; integrates environmental effects; potentially very sensitive | In general, does not indicate 'how much' water to apply; calibration required to determine 'control thresholds'; still largely at research/ development stage; little used for routine agronomy (except for thermal sensing in some situations) |
| (a) Tissue water status | Often been argued that leaf water status is the most appropriate measure for many physiological processes (e.g. photosynthesis), but this argument is generally erroneous (as it ignores root–shoot signaling) | All measures are subject to homeostatic regulation (especially leaf water status), therefore not sensitive (isohydric plants); sensitive to environmental conditions which can lead to short-term fluctuations greater than treatment differences |
| (i) Visible wilting | Easy to detect | Not precise; yield reduction often occurs before visible symptoms; hard to automate |
| (ii) Pressure chamber ($\psi$) | Widely accepted reference technique; most useful if estimating stem water potential (SWP), using either bagged leaves or suckers | Slow and labor intensive (therefore expensive, especially for predawn measurements); unsuitable for automation |
| (iii) Psychrometer ($\psi$) | Valuable, thermodynamically based measure of water status; can be automated | Requires sophisticated equipment and high level of technical skill, yet still unreliable in the long term |
| (v) Pressure probe | Can measure the pressure component of water potential which is the driving force for xylem flow and much cell function (e.g. growth) | Only suitable for experimental or laboratory systems |

| (vi) Xylem cavitation | Can be sensitive to increasing water stress | Cavitation frequency depends on stress prehistory; cavitation–water status curve shows hysteresis, with most cavitations occurring during drying, so cannot indicate successful rehydration |
|---|---|---|
| (b) Physiological responses | Potentially more sensitive than measures of tissue (especially leaf) water status | Often require sophisticated or complex equipment; require calibration to determine 'control thresholds' |
| (i) Stomatal conductance | Generally a very sensitive response, except in some anisohydric species | Large leaf-to-leaf variation requires much replication for reliable data |
| – Porometer | Accurate: the benchmark for research studies | Labor intensive so not suitable for commercial application; not readily automated (though some attempts have been made) |
| – Thermal sensing | Can be used remotely; capable of scaling up to large areas of crop (especially with imaging); imaging effectively averages many leaves; simple thermometers cheap and portable; well suited for monitoring purposes | Canopy temperature is affected by environmental conditions as well as by stomatal aperture, so needs calibration (e.g. using wet and dry reference surfaces |

Table 1. A summary of the main classes of irrigation scheduling techniques, indicating the major advantages and disadvantages (from Jones, 2004). Reproduced with kind permission of the author and Oxford University Press.

## 2.1 Measuring soil moisture

### 2.1.1 Water potential or volumetric water content?

Soil (substrate) water content can be expressed either in terms of the energy status of the water in the soil (i.e. matric potential, kPa) or as the amount of water in the substrate (most commonly expressed on a volumetric basis; % or $m^3 \cdot m^{-3}$). Both methods have advantages and disadvantages. Soil/substrate matric potential indicates how easily water is available to plants (Lea-Cox et al, 2011), but it does not provide information on how much total water is present in the substrate. Conversely, volumetric water content indicates how much water is present in a substrate, but not if this water is extractable by plant roots. This is especially important for soilless substrates, since mixtures of different components means that substrates have very different water-holding capacities and moisture release curves (deBoodt and Verdonck, 1972). Sensors that estimate water content (e.g. capacitance and TDR-type sensors) tend to be more reliable than those sensors measuring water availability (tensiometers and psychrometers); (Jones, 2008; Murray et al, 2004). A major disadvantage of almost all soil sensors, however, is their limited capability to measure soil moisture heterogeneity in the root zone, since they typically only sense a small volume around the sensor. Variation in soil water availability is well known, primarily as a function of variation in soil type, soil compaction and depth, among many sources of variation (e.g. organic matter content, porosity and rockiness). The use of large sensor arrays which may be necessary to get good representative readings of soil moisture tends to be limited by cost, but this could be overcome by sensor placement strategies.

Soilless substrates are used by the nursery and greenhouse industry for a multitude of reasons, primarily to reduce the incidence of soil-borne pathogens, increase root growth, and reduce labor, shipping and overall costs to the producer (Majsztrik et al., 2011). Over the years, many studies have shown large differences between soil and soilless substrates in the availability of water to root systems (Bunt, 1961; deBoodt and Verdonck, 1972). Soilless substrates, which in most cases have larger particle sizes and porosity, tend to release more water at very low matric potentials ($\Psi$m=-1 to -40 kPa) which is 10 to 100 times lower than similar plant-available water tensions in soils (Lea-Cox et al., 2011). Plant-available water (PAW) is the amount of water accessible to the plant, which is affected by the physical properties of the substrate, the geometry (height and width) of the container and the total volume of the container (Handreck and Black, 2002). Container root systems are usually confined within a short time after transplanting, and shoot : root ratios are usually larger than those of soil-grown plants, for similarly-aged plants. For all these reasons, maintaining the optimal water status of soilless substrates has been recognized as being critical for continued growth, not only because of limited water-holding capacity, but also because of the inadequacies of being able to accurately judge when plants require water (Karlovich & Fonteno, 1986). Although it is likely that mature plant root systems can extract substrate moisture at $\Psi$m less than -40 kPa, Leith and Burger (1989) and Kiehl et al. (1992) found significant growth reductions at substrate $\Psi$m as small as -16 kPa (0.16 Bar). This has major implications for choosing appropriate sensors for use in soilless substrates (see next section), as well as the measurement and automatic control of irrigation in these substrates.

### 2.1.2 Types of soil moisture sensors

Jones (2004) noted the various types of soil moisture sensors available at that time. The variety of soil moisture sensors (tensiometric, neutron, resistance, heat dissipation, psychrometric or dielectric) has continually evolved since then; the choices are now overwhelming, since each sensor may have specific strengths and weaknesses in a specific situation. Tensiometers have long been used to measure matric potential in soils (Smajstrla & Harrison, 1998) and in soilless substrates (Burger and Paul, 1987). Although tensiometers have proven to be valuable research tools, they have not been adopted widely in greenhouse and nursery production, mainly because of the problems with using them in highly porous soilless substrates. Tensiometers rely on direct contact between the porous ceramic tip and substrate moisture. If the substrate shrinks, or the tensiometer is disturbed, this contact may be disrupted. Air then enters and breaks the water column in the tensiometer, resulting in incorrect readings and maintenance issues (Zazueta et al., 1994). A number of next-generation soil moisture sensors have become available in the past decade from various manufacturers., e.g. Theta probe and SM200 (Delta T, Burwell, UK) and EC5, 5TM and 10HS sensors (Decagon Devices Inc., Pullman, WA, USA) which provide precise data in a wide range of soilless substrates. These sensors determine the volumetric moisture content by measuring the apparent dielectric constant of the soil or substrate. These sensors are easy to use and provide highly reproducible data (van Iersel et al., 2011). The Decagon range of sensors are designed to be installed in soils or substrates for longer periods of time and all interface with Decagon's range of EM50 nodes, datastation and Datatrac software (http://decagon.com/products). Dielectric sensors generally require substrate-specific calibrations, because the dielectric properties of different soils and substrates differ, affecting sensor output. The conversion between water potential and volumetric water content (VWC) varies substantially with soil type. It is possible to inter-convert matric potential to volumetric water content (Lea-Cox et al, 2011) for various sensors using substrate moisture release curves. However, such release curves are substrate-specific, and may change over time as the physical properties (e.g. pore size distribution) of the substrate changes (van Iersel et al., 2011) or root systems become more established. Fortunately, in most irrigation scheduling applications, the objective is simply to apply a volume of water that returns the soil moisture content to its original well-watered state. Changes in this total water-holding capacity (i.e. the maximum VWC reading) can easily be monitored for changes over time, i.e. after significant rainfall events or by periodically saturating the container with the embedded sensor.

More recently, hybrid 'tensiometer-like' sensors have been developed which use the principle of dielectric sensors to determine the water potential of substrates (e.g., Equitensiometer, Delta T; MPS-1, Decagon Devices) (van Iersel et al., 2011). An advantage of such sensors is that they do not require substrate-specific calibrations, since they measure the water content of the ceramic material, not that of the surrounding soil or substrate. Unfortunately, the sensors that are currently available are not very sensitive in the matric potential range where soilless substrates hold most plant-available water (0 to -10 kPa; deBoodt and Verdonck, 1972). In addition, it is not clear whether these sensors respond quickly enough to capture the rapid changes that can occur in soilless substrates (van Iersel et al., 2011).

### 2.2 Measuring plant water status

Automated irrigation techniques based on sensing plant water status are mostly in the developmental stage, in large part because of the variability of sensor readings and the lack of rugged sensors and reliable automated techniques (Table 1). It is usually necessary to supplement indicators of plant stress with additional information, such as crop evaporative demand (Jones, 2008); it is also hard to scale up these automated systems for many horticultural applications, since a detailed knowledge of crop development is required. Plant water use (transpiration) is a key process in the hydrologic cycle, and because photosynthetic uptake of $CO_2$ and transpiration are both controlled by stomata, it is strongly linked to plant productivity (Jones & Tardieu, 1998). Models that can accurately predict transpiration therefore have important applications for irrigation scheduling and crop yield. However, previous evidence (Jones, 2004) suggests that leaf water status is not the most useful indicator of plant water stress, and cannot therefore be used as the primary indicator of irrigation need as has sometimes been suggested. In fact leaf water status depends on a complex interaction of soil water availability and environmental and physiological factors (Jones, 1990). It is now clear that in some situations soil water status is sensed by the roots and this information is signaled to the shoots, perhaps by means of hydraulic signals (Christmann et al., 2009) and chemical messengers such as abscisic acid (Kim & van Iersel, 2011). Another general limitation to plant-based methods is that they do not usually give information on 'how much' irrigation to apply at any time, only whether irrigation is needed or not. None of the plant-based methods illustrated in Table 1 are well-adapted for automatic irrigation scheduling or control because of the difficulties measuring each variable (Jones, 2008). Typically, the use of any plant-based indicator for irrigation scheduling requires the definition of reference or threshold values, beyond which irrigation is necessary. Such threshold values are commonly determined for plants growing under non-limiting soil water supply (Fereres and Goldhamer, 2003), but obtaining extensive information on the behavior of these reference values as environmental conditions change will be an important stage in the development and validation of such methods.

## 2.3 Hybridizing sensing and modeling techniques for precision irrigation scheduling

Water budget calculations are relatively easy to use in scheduling irrigations, since there are simple algorithms available to calculate crop $E_T$ (typically using Penman-Monteith or other methods) that use local meteorological station or pan evaporation data (Fereres et al., 2003). All methods are based on calculating a reference $E_T$ that is multiplied by an empirically-determined crop coefficient (Kc) for each crop. At present there are good estimates of Kc values for many horticultural crops, even though most research has been conducted on the major field crops (Allen et al, 1998). However, there are virtually no $K_C$ values for ornamental species and most estimates of woody perennial crop water use are quite variable. Inaccuracies in Kc values can result in large potential errors in estimated soil moisture contents (Allen et al., 1998). The approach therefore works best where it is combined with regular soil moisture monitoring techniques that can help reset the model (e.g. after rainfall). A particular strength of book-keeping and volumetric soil-based approaches is that they not only address scheduling issues about "when to irrigate" but also about "how much to apply". Although useful for soil-based irrigation scheduling, there may be limitations on how quickly these calculations can be manually performed. This is especially important for greenhouse and container-nursery operations who may be cyclic irrigating containerized plants from 4-8 times per day (Tyler et al., 1996) to maintain available water in the root zone on hot, sunny or windy days.

Previous studies with a variety of crop, ornamental and turf species have reported that the use of appropriate scheduling methods and precision irrigation technologies can save a significant amount of water, while maintaining or increasing yield and product quality (Bacci et al., 2008; Beeson & Brooks, 2008; Blonquist et al., 2006; Fereres et al., 2003). Many of these empirical approaches have successfully incorporated environmental variables into various models, to further increase the precision of irrigation scheduling (e.g. Treder et al., 1997). It is imperative that we connect our capability for precision water applications with a knowledge of real-time plant water use. We need to improve our ability to predict plant water use in real-time using various technologies. As an example of this approach, van Iersel and his group have shown with various studies (Burnett and van Iersel, 2008; Kim and van Iersel, 2009; Nemali and van Iersel, 2006; van Iersel et al., 2009; 2010; 2011) that automated irrigation using soil moisture sensors allows for the very precise irrigation of greenhouse crops in soilless substrates. In addition, they maintained very low substrate moisture contents at very precise levels which advances our capability to use precision irrigation scheduling for  regulated deficit irrigation (RDI) techniques (Jones, 2004), to increase fruit crop quality (Fereres et al., 2003), and aid in precision nutrient (Lea-Cox et al, 2001; Ristvey et al, 2004) and disease management (Lea-Cox et al, 2006).

Most recently, Kim and van Iersel (2011) have demonstrated that the measured daily evapotranspiration of petunia in the greenhouse can be accurately modeled with measurements of crop growth (days after planting, DAP), daily light integral (DLI), vapor pressure deficit (VPD) and air temperature. All these environmental fluxes obviously affect transpiration on a continuous basis. Ambient light affects plant water use due to its effects on evaporation and stomatal opening (Pieruschka et al., 2010). Vapor pressure deficit is the driving force for transpiration and also affects stomatal regulation (Taiz and Zeiger, 2006) while temperature affects $E_T$ and plant metabolic activity (Allen et al., 1998; van Iersel, 2003). The importance of Kim & van Iersel's empirical modeling approach is how they have demonstrated the sensitivity of plant water use to these four easily-measured variables. Thus, with a few inexpensive sensors (temperature, relative humidity and photosynthetic photon flux, PPF) and some simple software tools that can integrate these variables on short time-scales, it now appears possible to predict hourly plant water use for greenhouse crops with real precision. It should be noted however, that these models still require rigorous validation for production conditions.

However, for these types of models to work in an external environment, it is likely that the complexity of our predictive water use models will have to increase, to incorporate additional variables. Water use by perennial woody crop species is much more complicated due to external environmental conditions (for example how VPD and leaf temperature are affected by wind speed and boundary layer effects on canopies; LAI effects on PPF interception). For example, Bowden et al. (2005) outlined an automated sensor-based irrigation system for nurseries that could calculate plant water consumption from species and genotype-specific plant physiological responses. The MAESTRA [Multi-Array Evaporation Stand Tree Radiation A] model (Wang and Jarvis, 1990) is a three-dimensional process-based model that computes transpiration, photosynthesis, and absorbed radiation within individual tree crowns at relatively short time (15-minute) intervals. The model is described more fully by Bauerle & Bowden (2011b) and has been modified and previously validated to estimate deciduous tree transpiration (Bauerle et al., 2002; Bowden & Bauerle, 2008) and within-crown light interception (Bauerle et al., 2004). The model applies physiological equations to sub-volumes of the tree crown and then sums and/or averages the values for entire canopies. Additionally, species-specific physiological values can be incorporated into model calculations, potentially yielding more accurate estimates of whole tree transpiration. The model holds potential advantages for nursery, forest, and orchard water use prediction in that structural parameters such as tree position, crown shape, and tree dimensions are specified.

Bowden et al. (2005) briefly illustrated how the model estimates of water use and plant water requirements are outputted from MAESTRA and used to both make irrigation decisions (command executed by a sensor node) and visualize model updates via a graphic user interface (Bauerle et al., 2006). Within each 15-minute time step, the model adjusts transpiration based on interactions between environmental, soil moisture, and plant physiological response. The substrate moisture deficit calculation is described in Bauerle et al. (2002). An updated substrate moisture value is carried into the next time step for input into the substrate moisture deficit sub-routine. The calculated moisture deficit value is one of the input values required to calculate the amount of stomatal conductance regulation and hence, interacts with other equations to derive whole plant water use. Overall, this GUI (Bowden et al., 2005; Bauerle et al., 2006) provides a user friendly interface to a complex set of calculations. In this way, whole tree water use estimates can be rapidly visualized for either sensor node or human based irrigation decision management. Bauerle and his group are actively working to further refine the MAESTRA model for incorporation into the irrigation scheduling decision support system in our current project (Lea-Cox et al., 2010b).

## 3. Utilizing the power of sensor networks

### 3.1 Wireless Sensor Networks

A WSN is typically comprised of radio frequency transceivers, sensors, microcontrollers and power sources (Akyildiz et al., 2002). Recent advances in wireless sensor networking technology have led to the development of low cost, low power, multifunctional sensor nodes. These nodes can be clustered in close proximity to provide dense sensing capabilities, or deployed in a more distributed fashion (Fig. 1). We shall describe the commercially-available Decagon Devices WSN, since we have the most experience with that system, although there are other commercial companies that have similarly available irrigation and environmental WSN systems, e.g. Adcon Telemetry Int. (Klosterneuburg, Austria; http://adcon.at), Delta-T Devices (Cambridge, UK; http://delta-t.co.uk) and PureSense (Fresno, CA; http://puresense.com).

Figure 1 shows the type of WSN that we have deployed in multiple research and commercial sites. Whenever necessary, the accumulated data is transmitted from each of the sensor nodes in the production area using a 900 MHz radio card (although other companies use other frequencies), to a 'base' datastation connected to a personal computer on the farm. Incoming data is inputted into a software program (e.g. DataTrac v.3.2; Decagon Devices) that is installed on a low-cost computer. The software then plots and displays the sensor information from each of the nodes. Data is appended to existing data, so information can be graphically displayed over multiple time scales, depending on user preference. Alternatively, data from a field node can be transmitted directly to a server via the internet using a 3G wireless node (e.g. EM50G, Decagon Devices). The logged data is then accessed from the server over an internet website, using the same DataTrac software previously described. In this way, a grower can develop a scaleable network of sensors that allows for the monitoring of soil moisture and environmental data, in real time. The advantages of these WSN's are fairly obvious – they provide information at the "micro-scale" which can be expanded to any resolution, determined for a specific production operation, for specific needs. This system also provides a mechanism for local (i.e. a decision made locally by the node, based on local sensor readings /setpoints) or the global control (information relayed to the nodes from an external database) of irrigation scheduling (Fig. 1) , depending upon grower preferences and needs (Kohanbash et al, 2011). We are currently in the process of deploying and testing next-generation nodes with these various capabilities.

Any combination of environmental sensors, including soil moisture and electrical conductivity, soil and air temperature, relative humidity, anemometer (wind speed and direction), rain gauge and light (PPF and net radiation) sensors can be connected to the nodes, according to user needs. Decagon nodes collect data every minute, which is averaged and logged on a 1, 5- 15-min or greater time scale, according to required precision. Longer sampling times result in a considerable increase in battery life, but power consumption will also vary greatly with different systems. With Decagon EM50 nodes, a 15-min average setting typically results in > 12-month battery life from 5 'AA' batteries under normal temperature (-5 to 40°C) conditions (J.D. Lea-Cox, *pers. obs.*). Battery life is also affected by the number of times the field nodes are downloaded and the settings employed; typically nodes are downloaded 1-10 times a day.

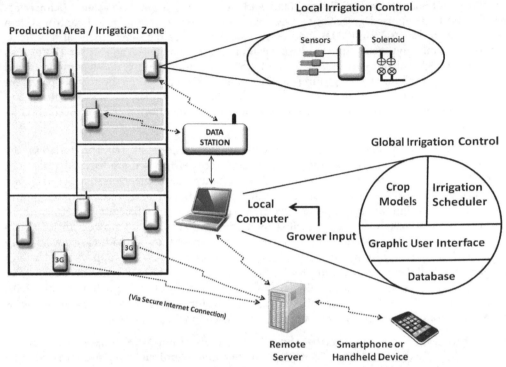

Fig. 1. A schematic of a farm-scale WSN for precision irrigation scheduling (adapted from Balendock, 2009), to illustrate networks deployed by our group (Lea-Cox et al., 2010b).

## 3.2 Scaleable, adaptable and reconfigurable capabilities

One of the most important features of these WSNs is that a grower can purchase a small network and scale up and/or reconfigure the sensor network to meet specific needs, over time. These networks can provide a fixed capability, but networks can be more fully utilized if a "nimble networks" concept is used. In this way, growers can move sensor nodes quickly and easily within the production area for shorter periods of time, to address current issues and problems, e.g. to address the water requirements of a specific indicator species in a drought, or to monitor irrigations to reduce the incidence of disease in a crop (Lea-Cox et al., 2006). This nimble network approach can more fully utilize WSN capabilities and is one of the most powerful ways of realizing a quick return on investment in equipment. We think that there are many situations where a grower could have a payback period for a small network within a single crop cycle, if the information is utilized for better irrigation and crop management decisions.

## 3.3 Wireless sensor network development for irrigation scheduling

A number of WSN's with various topologies (e.g. star, mesh-network) have been developed and investigated by different researchers in the past decade (Ruiz-Garcia et al., 2009), including WSN's for irrigation scheduling in cotton (Vellidis et al., 2008), center-pivot irrigation (O'Shaughnessy & Evett, 2008) and linear-move irrigation systems (Kim et al., 2008). The first reported greenhouse WSN was a bluetooth monitoring and control system developed by Liu & Ying (2003). Yoo et al. (2007) describes the deployment of a wireless environmental monitoring and control system in greenhouses; Wang et al. (2008) also developed a specialized wireless sensor node to monitor temperature, relative humidity and light inside greenhouses. Our group (Lea-Cox et al., 2007) reported on the early deployment of a WSN within a cut-flower greenhouse, where a number of soil moisture and environmental sensor nodes were deployed for real-time monitoring of crop production by the grower.

With regards to large on-farm WSN deployments, Balendonck et al., (2009) reported on the FLOW-AID project that has many of the same objectives that we are focused upon, i.e., providing growers with a safe, efficient and cost-effective management system for irrigation scheduling. The FLOW-AID project is integrating innovative monitoring and control technologies within an appropriate decision support system (Balendonck et al., 2007; Ferentinos et al., 2003) that is accessible over the internet, to assist growers in long-term farm zoning and crop planning. It is especially focused on providing growers with regulated deficit irrigation and soil salinity management tools. To support shorter-term irrigation scheduling, a scheduling tool is being developed which allocates available water among several plots and schedules irrigation for each plot (Stanghellini et al., 2007; Anastasiou et al., 2008). To assist this advanced scheduling tool, a crop response model is being developed and used to predict crop stress (Balendonck et al., 2009).

We outlined the major engineering and scientific goals of our WSN project earlier in this chapter (Lea-Cox et al., 2010a). To explain further, this interdisciplinary project is taking a commercially-available WSN product (Decagon Devices, Inc.) and retooling it to support the irrigation scheduling requirements of field nurseries, container nurseries, greenhouse operations and green roof systems, as analogs for many intensive agricultural production and environmental management systems. Our global goals are to develop a more integrative and mechanistic understanding of plant water requirements, to more precisely schedule irrigation events with WSN technology. We are working across various scales of production, using small and large commercial farms which allow us to take a systems approach to defining the hardware and software required to meet the needs of these highly intensive specialty crop systems. In addition to the ornamental industry, there are many parallel needs that we are addressing for WSN adoption by field-grown fruit, nut and berry production, as well as field and greenhouse vegetable production. As part of the project, economic, environmental and social analyses will identify costs and benefits of WSN technology to the industry and society, including barriers to adoption. The project directly involves commercial growers throughout the process, using deployments in commercial operations as test sites. This will help ensure product satisfaction of the next generation of hardware and software developed by our various teams (Lea-Cox et al., 2010b; http://www.smart-farms.net). Each farm and research test site is instrumented with a sensor network(s) to provide real-time environmental data for scientific and technological development. Data streams are monitored on a day-to-day basis by growers, engineers and scientists, which drives a daily dialogue between the growers and various working groups.

The role of the engineering team is to develop, deploy and maintain the next generation of wireless sensor networks (Fig. 1). Their major task is to develop the hardware and software capable of supporting advanced monitoring and control of irrigation scheduling, implementing a hybrid sensor and modeling approach. A major focus of this effort is the development of advanced software which will provide advanced user control, in addition to database filtering and analysis. The software will refine incoming data and provide an easy-to-use computer program for a non-expert user to easily visualize the information from the WSN, and schedule irrigation events based on user preference, or utilizing automatic (set-point) control. The scientific modeling group (Bauerle et al., 2011a; Kim and van Iersel, 2011; Starry et al., 2011) are developing and validating the various models, which form the basis of the species- and environmental-specific software. These models interface with the WSN database via an open application programming interface, which integrates the models with the irrigation scheduling monitoring and control functions. This will enable more predictive (feed-forward) management of water use, based upon the underlying plant and environmental water-use models.

The role of the scientific and extension teams is to ensure that the precision and accuracy of the data gathered (and hence the quality of the models incorporated in the decision support software) are of the highest possible quality and reliability. There are a number of critical research objectives that span the various production environments: (a) characterize the spatial and temporal variability of environmental parameters in both root and shoot canopies, since we need to place sensors for maximum precision and economic benefit; (b) characterize sensor performance and precision, so we match the right sensor with the right application; (c) integrate the knowledge gained from (a) and (b), to ensure that the irrigation scheduling decisions made (either manually or automatically) satisfy plant water requirements in real-time, while placing a minimum burden on the grower for managing the system. We elaborate further on some of these critical objectives below in section 3.4. However, our primary project objective is to provide a cost-effective WSN that provides quality data for minimal cost to growers, both small and large. Our grower's production areas range from 0.5 to over 250 ha in extent, with multiple irrigation zones / crop species. To that end, our economic and environmental analysis team members are gathering specific economic, resource use and environmental data from each production site through a series of on-farm visits and assessments. Larger outreach (survey) efforts across the United States will validate results from our intensive economic analysis of the commercial operations in the study. Some early WSN deployment strategies and results from the project are illustrated later in this chapter. Further project information, results and learning modules are available from our interactive website at http://www.smart-farms.net.

## 3.4 Sensor network deployment issues and strategies

### 3.4.1 Spatial and temporal variability

Understanding spatial and temporal variability of environmental data is one of the most important aspects of deploying WSNs in any real-world application, since these dynamics not only determine the appropriate position of a sensor, but the precision of the sensor data is of course greatly affected by the immediate environment and the forces acting on that environment. This is the realm of environmental biophysics (Campbell and Norman, 1998; Jones, 1992) and environmental plant physiology (Nobel, 2009) which forms the basis of our efforts to sense and model the environment.

Figure 2 illustrates soil moisture variability from 10HS soil moisture sensors at two depths (15 and 30cm below the soil surface) in five replicate *Acer rubrum* (Red maple) trees from May through Sept., 2009 (Lea-Cox, unpublished data). Sensors were calibrated to the specific soil type found on this farm (Lea-Cox, Black, Ristvey & Ross, 2008).

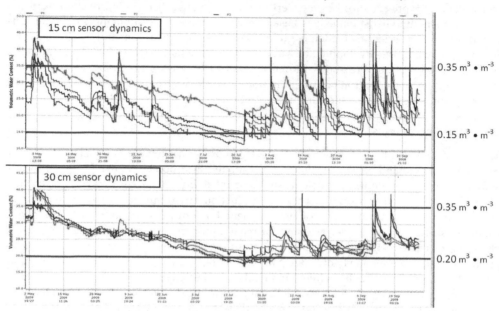

Fig. 2. Soil moisture dynamics at 15cm and 30cm depths in the root zone of five replicate *Acer rubrum* trees from May through September, 2009. Stacked data from two nodes is graphically displayed using an earlier version of DataTrac (v. 2.78, Decagon Devices).

As can be seen, 15cm data were more variable throughout the entire season (Fig. 2). Trees were irrigated for 1-2 hours with drip irrigation on a daily basis throughout most of the study (Lea-Cox, Black, Ristvey & Ross, 2008), except after major rainfall events restored the soil water contents above 0.25 m³ • m⁻³ (Fig. 2). Changes in daily water content (tree uptake) are evident immediately after these rainfall events, particularly in the 15cm dynamics. Soil moisture dynamics at the 30cm depth were much less variable between trees at all times during the season, and soil moisture at this depth did not fall below 0.20 m³ • m⁻³ during this year, despite relatively low rainfall totals during the summer (data not shown).

Figure 3 shows similar soil moisture data from 10HS sensors at 15cm depth from *Cornus florida* trees, but these trees were grown in a pine bark soilless substrate in 56-liter containers in a container-nursery operation. Firstly, note that the average substrate moisture is around 0.5 m³ • m⁻³, since this organic substrate has a high water-holding capacity and also because this grower typically irrigates 2-3 times per day, with small low-volume microsprinkler events (1.5L in 6 minutes) during summer months.

Note also how quickly substrate moisture decreases with plant uptake when morning and early afternoon irrigation events are skipped, due to the relatively low amounts of total water in the container (Fig. 3). Note also that real-time irrigation applications per tree are easily measured using a small tipping rain gauge with a rain cover, with an additional microsprinkler head inserted under the cover (Lea-Cox et al., 2010b). The volumes displayed (Fig. 3) give the grower instantaneous feedback and tie soil moisture contents directly to irrigation events and the volumes applied. We are using the same tipping rain gauges to give leaching volumes from pot-in-pot containers with an underground drainage system, to provide approximate daily irrigation water budgets (i.e. Irrigation + Rainfall - Leaching = $\delta$VWC $\approx E_T$) for additional indicator species on the farm (Lea-Cox et al., 2010b).

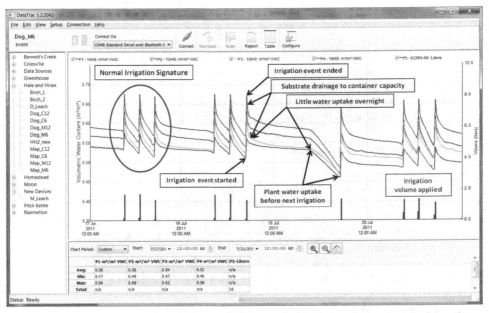

Fig. 3. Typical container moisture dynamics before and after irrigation events in four *Cornus florida* trees. Data is graphically displayed using the most recent version of DataTrac (v. 3.2).

### 3.4.2 Sensor placement

Changes in substrate VWC due to daily water use of a crop can be used to control irrigation events, but placement of sensors is very important in container production, because of the non-uniform distribution of water within a soilless substrate and container. van Iersel et al. (2011) illustrated this point, by calculating the *rate of change* in substrate VWC. They noted that the maximum rate of decrease in VWC occurred at the bottom of the container in a greenhouse study and closely followed changes in solar radiation, suggesting that changes in VWC were driven by root water uptake from the lower part of the container. However, the vertical gradient in substrate VWC also changed over time i.e., the VWC of the bottom layer decreased much more rapidly than that of the upper layers, likely because of the root distribution within the container. Apparently, the lack of roots in the upper part of the substrate resulted in little water uptake from that substrate layer, and vertical water movement in the container was not fast enough to prevent the middle layers from getting drier than the upper layer. If these findings can be generalized for other container-grown species, it would greatly increase our understanding for correct sensor placement in root zones, simplifying placement and increasing the precision of information for controlling irrigation events. However, since root distribution is affected by irrigation method, optimal placement of soil moisture sensors for irrigation control may depend on how the crop is irrigated (van Iersel et al., 2011).

Sensor placement is especially challenging for crops grown in large containers over relatively long periods of time. Barnard et al. (2011) examined the spatial and temporal variation in VWC among 10 tree species in large containers in a container nursery, and found significant differences within containers and among species. Based on their initial results, they recommended species-specific sensor deployment. For such crops, where root distribution within the container may change dramatically throughout the production period, it may be necessary to move the soil moisture sensor as root distribution changes, or it may be possible to use a soil moisture sensor that can sense the substrate water content throughout most of the container. It is therefore likely that a hybrid sensor and crop water use model approach will have greater degree of precision for automated irrigation scheduling, a feature desired by many greenhouse and container-nursery growers.

### 3.4.3 Using indicator plant species

For many ornamental operations, it is unlikely that we will be able to sense the water needs of all crop species being grown. Many growers however are familiar with the concept of using indicator species (i.e., species that have high and low water use, on average), which are used to inform irrigation schedules for similar types of plants (Yeager et al., 2007). For this reason, we are developing crop models which include a number of these indicator species in the decision support software. Part of this strategy is also to engage the larger research community in the development and incorporation of additional specific crop models (e.g. Warsaw et al., 2009) in future irrigation decision support systems.

### 3.4.4 Microclimatic data

The gathering and seamless integration of real-time environmental data is integral to the development and implementation of crop-specific (Bauerle et al., 2010; Kim and van Iersel, 2011) and environmental models (Starry, 2011). Typical microclimatic data which is gathered by "weather" nodes is displayed in Fig. 4. Tools within DataTrac v.3.2 now allow for the calculation and plotting of integrated data, such as vapor pressure deficit, daily light integral and accumulated degree days, as simple derivatives of this instantaneous data. Apart from the integration of this data into various crop, environmental and disease development models, this microclimatic data has many other direct practical benefits for producers, e .g. the use of real-time T/RH and wind speed data for precision timing of spray schedules in the field. Longer-term seasonal information for light, precipitation and maximum/minimum air and soil temperatures are very informative for growers to assess crop growth development and other production variables e.g. residual soil nutrient values.

### 3.4.5 Predictive irrigation scheduling

The integration of real-time microclimatic data into crop-specific and environmental water use models is the next step in our development path; we have successfully parameterized petunia (Kim and van Iersel, 2011), red maple (Bauerle et al., 2011b) and green roof

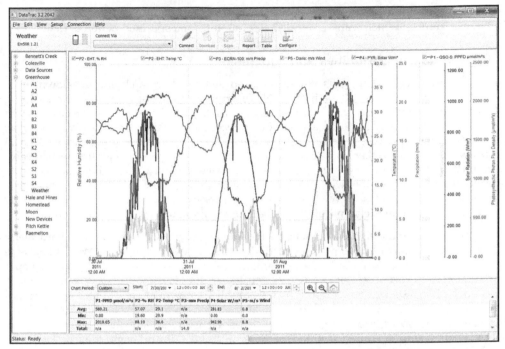

Fig. 4. Weather data from a sensor node with a typical suite of environmental sensors, including total radiation (black), photosynthetic photon flux (purple), relative humidity, RH (blue), air temperature, T (red) and wind speed (green).

stormwater runoff (Starry et al, 2011) models, and we are verifying and validating those models with current research projects. Our modeling and engineering teams are interfacing these models with a testbed sensorweb system (Kohanbash et al., 2011). In their paper, they present a framework for integrating physiological models into WSN for advanced irrigation scheduling. They note that the ability to gather high resolution data, interpret it, and create an actionable conclusion is a critical ability for a WSN.

Kohanbash et al. (2011) outlined our irrigation scheduling programming logic where growers create a schedule for when water needs to be applied, and then the schedule is interrupted as needed. This system provides growers with four operating modes: (1) a schedule-based controller very similar to what is commonly used in the industry. Within the schedule, there are two different options to over-ride the schedule to decrease the irrigation time (a) a local setpoint controller and (b) a global controller. The schedule + local setpoint controller enables the sensor node to make local control decisions based on sensors attached to the node. The schedule + global controller allows the grower to use data from *any* node in the network, calculated data or model data to control the irrigation and consequently determine if the schedule should be interrupted. The fourth mode is a manual override mode that allows the grower to water in traditional mode, for a given number of minutes. This irrigation scheduling flexibility gives a grower the ability to control how water gets applied to an irrigation zone, with various user-defined parameters. The user can choose between a mode where water will be applied slowly with small delays between irrigation events to allow water to reach subsurface sensors (micro-pulse irrigation; Lea-Cox et al., 2009) or a mode in which water is applied continuously for a specified period of time.

## 4. Challenges, opportunities and conclusions

Of course there are many areas where we need additional research and development, to provide the maximum cost benefit of WSNs for growers. Challenges include standardizing WSN protocols and communication frequencies, as they can be confusing for growers and researchers alike. Nodes operating at lower frequencies (900 MHz) typically have an increased range and can penetrate tree canopies better than higher frequency (2.4 GHz) nodes with reduced packet loss. Battery-operated nodes are typical; integrating rechargeable capabilities into sensor nodes is important, especially if control capabilities are going to become standard, since this will greatly increase power requirements. Another challenge is working with large datasets. We have to educate ourselves as to the resolution required for optimum precision in each environment, keeping the ultimate use of the data in mind.

The maintenance and calibration of sensors and equipment is an ongoing concern, particularly for growers who may be uncomfortable with the technology and equipment. We definitely see an opportunity for paid consultants to maintain and remotely monitor WSNs for optimum performance. As part of our project, we are developing an online knowledge center, to provide assistance and guidance about various aspects of WSN deployment, sensor use, strategies and best practices. We need to integrate better data analysis tools to handle large volumes of data from sensor networks. We also need to do a thorough user interface study on how growers actually use computer interfaces and to determine what features are needed. Predictive models for plant water use, environmental and disease management tools are rapidly being developed for growers, but we need to validate and verify these models for use in different environments. Incorporation of models into WSNs for decision-making appears to be relatively easy, but there are many details which have yet to be worked out.

There are many layers to the socio-economic analysis our economic team is performing. Of course there are many direct benefits of precision irrigation scheduling that can be accrued by the grower, such as saving on water, labor, electricity, and fertilizer costs. However, there are many indirect (e.g. reduced disease incidence, fungicide costs) and societal benefits (reduced nutrient runoff, groundwater consumption) that may have much larger benefits over the long-term for all agricultural producers. Most importantly, we need to quantify the return on investment that a grower could expect to achieve, and to be able to scale those benefits for small producers, along with scaling WSN deployments. We are also interested in documenting perceived and real barriers to adoption. Our socio-economic team is actively surveying a large number of growers with a detailed survey, to compare the use of sensor technology and irrigation decisions by early and late adopters.

In conclusion, we believe that there have been some real advances in WSNs for precision irrigation scheduling in recent years. Of course many challenges still remain, but we believe that WSNs are a fast-maturing technology that will be rapidly adopted by many growers in the near future.

## 5. Acknowledgements

We gratefully acknowledge grant support from the USDA-NIFA Specialty Crops Research Initiative Award # 2009-51181-05768. I would like to thank Bill Bauerle, Marc van Iersel,

David Kohanbash and Bruk Belayneh for their excellent reviews which improved this Chapter. I would also like to personally acknowledge the dedication and talent of everyone on this project, without whom none of this research and development would be possible.

## 6. References

Anastasiou, A., Savvas, D., Pasgianos, G., Stangellini. C., Kempkes, F. & Sigrimis, N. 2008. Decision Support for Optimised Irrigation Scheduling. *Acta Hort.* 807:253–258.

Agarwal, A., delos Angeles, M.S., Bhatia, R., Cheret, I., Davila-Poblete, S., Falkenmark, M., Gonzalez Villarreal, F., Jønch-Clausen, T., Kadi, M.A., Kindler, J., Rees, J., Roberts, P., Rogers, P., Solanes, M. & Wright. A. 2000. Integrated Water Resources Management. *Global Water Partnership.* Denmark.

Akyildiz, I.F., Su, W., Sankarasubramaniam, Y. & Cayirci, E. 2002. Wireless Sensor Networks: A Survey. *Computer Networks* 38:393-422.

Allen, R. G., Pereira, L. S., Raes, D. & Smith. M. 1998. Crop Evapotranspiration: Guidelines for Computing Crop Water Requirements. *FAO Irrigation and drainage paper no.56.* FAO, Rome, Italy.

Bacci, L., Battista, P. and Rapi, B. 2008. An Integrated Method for Irrigation Scheduling of Potted Plants. *Sci. Hortic.* 116:89-97.

Bauerle, W.L., Post, C.J., McLeod, M.E., Dudley J.B. & Toler, J.E. 2002. Measurement and Modeling of the Transpiration of a Temperate Red Maple Container Nursery. *Agric. For. Meteorol.* 114:45-57.

Bauerle, W.L., Bowden, J.D., McLeod M.F. & Toler, J.E. 2004. Modeling Intracrown and Intracanopy Interactions in Red Maple: Assessment of light transfer on carbon dioxide and water vapor exchange. *Tree Physiol.* 24:589-597.

Bauerle, W.L., Timlin, D.J., Pachepsky, Ya.A. & Anantharamu., S. 2006. Adaptation of the Biological Simulation Model MAESTRA for Use in a Generic User Interface. *Agron. J.* 98:220-228.

Bauerle, W.L. & Bowden, J.D. 2011a. Predicting Transpiration Response to Climate Change: Insights on Physiological and Morphological Interactions that Modulate Water Exchange from Leaves to Canopies. *HortScience* 46:163-166.

Bauerle, W.L. & Bowden, J.D. 2011b. Separating Foliar Physiology from Morphology Reveals the Relative Roles of Vertically Structured Transpiration Factors Within Red Maple Crowns and Limitations of Larger Scale Models. *J. Exp. Bot.* 62:4295-4307.

Balendonck, J., Stanghellini, C., & Hemming, J. 2007. Farm Level Optimal Water Management: Assistant for Irrigation under Deficit (FLOW-AID). *In: Water Saving in Mediterranean Agriculture & Future Research Needs,* Bari, Italy, Lamaddalena N. & Bogliotti, C. (Eds.), CIHEAM, Bari, Italy. pp. 301-312.

Balendonck, J., Stanghellini, C., Hemming, J., Kempkes F.L.K. & van Tuijl, B.A.J. 2009. Farm Level Optimal Water Management: Assistant for Irrigation under Deficit (FLOW-AID). *Acta Hort* 807: 247-252.

Barnard, D.M, Daniels, A.B. & Bauerle, W.L., 2011. Optimizing Substrate Moisture Measurements in Containerized Nurseries: Insights on Spatial and Temporal Variability. *HortScience* 46(9): S207. (Abstr.)

Barrenetxea, G., Ingelrest, F., Schaefer, G., & Vetterli, M. 2008. The Hitchhiker's Guide to Successful Wireless Sensor Network Deployments. *In: Proc. 6th ACM Int. Conf. Embedded Networked Sensor Systems (SenSys)* Raleigh, NC, Nov. 5 -7, 2008. pp. 43-56.

Beeson Jr., R.C. & Brooks. J. 2008. Evaluation of a Model Based and Reference Crop Evapotranspiration (ETo) for Precision Irrigation using Overhead Sprinklers during Nursery Production of *Ligustrum japonica. Acta Hort.* 792:85-90.

Bitelli, M. 2011. Measuring Soil Water Content: A Review. *HortTechnology* 21: 293-300.

Blonquist, J.J.M., Jones, S.B. & Robinson. D.A. 2006. Precise Irrigation Scheduling for Turfgrass using a Subsurface Electromagnetic Soil Moisture Sensor. *Agricultural Water Management* 84:153-165.

Bowden, J.D. & Bauerle, W.L. 2008. Measuring and Modeling the Variation in Species-Specific Transpiration in Temperate Deciduous Hardwoods. *Tree Physiol.* 28:1675-1683.

Bowden, J. D., Bauerle, W.L., Lea-Cox, J.D. & Kantor, G.F. 2005. Irrigation Scheduling: An Overview of the Potential to Integrate Modeling and Sensing Techniques in a Windows-based Environment. *Proc. Southern Nursery Assoc. Res. Conf.* 50:577-579.

Bunt, A.C. 1961. Some Physical Properties of Pot-Plant Composts and their Effect on Plant Growth. *Plant and Soil* 13:322-332.

Burger, D.W. & Paul, J.L. 1987. Soil Moisture Measurements in Containers with Solid State, Electronic Tensiometers. *HortScience* 22:309-310.

Burnett, S.E. & van Iersel, M.W. 2008. Morphology and Irrigation Efficiency of *Gaura Lindheimeri* Grown with Capacitance-Sensor Controlled Irrigation. *HortScience* 43:1555-1560.

Campbell, G. S. & Norman. J. 1998. An Introduction to Environmental Biophysics. 2nd Ed. *Springer* New York, NY.

Cifre, J., Bota, J., Escalona, J.M., Medrano, H. & Flexas, J. 2005. Physiological Tools for Irrigation Scheduling in Grapevine (*Vitis vinifera* L.). *Agric. Ecosys. Environ.* 106:159-170.

Christmann, A., Weiler, E.W., Steudle, E. & Grill. E. 2007. A Hydraulic Signal in Root-to-Shoot Signaling of Water Shortage. *Plant J.* 52:167-174.

Dane, J.H. & Topp, G.C. 2002. Methods of Soil Analysis. Part 4: Physical Methods. *Amer. Soc. Soil Sci.*, Madison, WI.

deBoodt, M. & O. Verdonck. 1972. The Physical Properties of Substrates in Horticulture. *Acta Hort.* 26:37-44.

Ferentinos, K.P., Anastasiou, A., Pasgianos, G.D., Arvanitis, K.G. & Sigrimis, N. 2003. A DSS as a Tool to Optimal Water Management in Soilless Cultures under Saline Conditions. *Acta Hort.* 609:289-296.

Fereres, E.; Goldhamer, D.A. & Parsons, L.R. 2003. Irrigation Water Management of Horticultural Crops. *HortScience* 38:1036-1042.

Fereres, E. & Goldhamer, D.A. 2003. Suitability of Stem Diameter Variations and Water Potential as Indicators for Irrigation Scheduling of Almond Trees. *J. Hort. Sci Biotech.* 78:139-144.

Fry, A., 2005. Facts and Trends, Water. *World Business Council for Sustainable Development, Earthprint Ltd.* ISBN 2-940240-70-1.

Guimerà, J. 1998. Anomalously High Nitrate Concentrations in Ground Water. *GroundWater* 36:275-282.

Handreck, K. & Black. N. 2002. Growing Media for Ornamental Plants and Turf (3rd Ed.). *Univ. New South Wales Press*, Sydney, Australia.

Hutson, S.S., Barber, N.L., Kenny, J.F., Linsey, K.S., Lumia, D.S. & Maupin, M.A. 2004. Estimated Use of Water in the United States in 2000. *Circular* 1268. *U.S. Geological Survey*, U.S. Dept. Interior. Reston, VA.

Jones, H.G. 1992. Plants and Microclimate: A Quantitative Approach to Environmental Plant Physiology. (2nd ed.) *Cambridge University Press*. Cambridge. UK.

Jones, H.G. 1990. Physiological-Aspects of the Control of Water Status in Horticultural Crops. *HortScience* 25: 19–26.

Jones, H.G. 2004. Irrigation Scheduling: Advantages and Pitfalls of Plant-Based Methods. J. Exp. Bot. 55:2427-2436.

Jones, H.G. 2008. Irrigation Scheduling - Comparison of Soil, Plant and Atmosphere Monitoring Approaches. *Acta Hort.* 792: 391-403.

Jones, H.G. & Tardieu F. 1998. Modeling Water Relations of Horticultural Crops: A Review. *Scientia Horticulturae.* 74:21–46.

Karlovich, P.T. & Fonteno, W.C. 1986. The Effect of Soil Moisture Tension and Volume Moisture on the Growth of *Chrysanthemum* in Three Container Media. *J. Amer. Soc. Hort. Sci.* 111:191-195.

Kiehl, P.A., Liel, J.H. & Buerger, D.W. 1992. Growth Response of *Chrysanthemum* to Various Container Medium Moisture Tensions Levels. *J. Amer. Soc. Hort. Sci.* 117:224-229.

Kim, Y., Evans, R.G. & Iversen, W.M. 2008. Remote Sensing and Control of an Irrigation System using a Distributed Wireless Sensor Network. *IEEE Trans. Instrum. Meas.* 57:1379-1387.

Kim, J. & van Iersel, M.W. 2009. Daily Water Use of Abutilon and Lantana at Various Substrate Water Contents. *Proc. Southern Nursery Assn. Res. Conf.* 54:12-16.

Kim, J., van Iersel, M.W. & Burnett, S.E. 2011. Estimating Daily Water Use of Two Petunia Cultivars Based on Plant and Environmental Factors. *HortScience* 46:1287-1293.

Kim, J. & van Iersel. M.W. 2011. Abscisic Acid Drenches can Reduce Water Use and Extend Shelf Life of *Salvia splendens. Sci. Hort.* 127: 420–423.

Kohanbash D., Valada, A. & Kantor, G.F. 2011. Wireless Sensor Networks and Actionable Modeling for Intelligent Irrigation. *Amer. Soc. Agric. Biol. Eng.* 7-12th August, 2011. Louisville, KY. Paper #1111174. 7p.

Lea-Cox, J.D., Ross, D.S. and Teffeau, K.M. 2001. A Water And Nutrient Management Planning Process For Container Nursery And Greenhouse Production Systems In Maryland. *J. Environ. Hort.* 19:230-236.

Lea-Cox, J. D., Arguedas-Rodriguez, F. R., Amador, P., Quesada, G. & Mendez, C.H. 2006. Management of the Water Status of a Gravel Substrate by Ech20 Probes to Reduce *Rhizopus* Incidence in the Container Production of *Kalanchoe blossfeldiana. Proc. Southern Nursery Assoc. Res. Conf.* 51: 511-517.

Lea-Cox, J. D., Kantor, G. F., Anhalt, J., Ristvey A.G. & Ross. D.S. 2007. A Wireless Sensor Network for the Nursery and Greenhouse Industry. *Proc. Southern Nursery Assoc. Res. Conf.* 52: 454-458.

Lea-Cox, J.D., Ristvey, A.G. Arguedas-Rodriguez, F.R., Ross, D.S., Anhalt J. & Kantor. G.F. 2008. A Low-cost Multihop Wireless Sensor Network, Enabling Real-Time Management of Environmental Data for the Greenhouse and Nursery Industry. *Acta Hort.* 801:523-529.

Lea-Cox, J.D., Black, S., Ristvey A.G. & Ross. D.S. 2008. Towards Precision Scheduling of Water and Nutrient Applications, Utilizing a Wireless Sensor Network on an Ornamental Tree Farm. *Proc. Southern Nursery Assoc. Res. Conf.* 53: 32-37.

Lea-Cox, J.D., Ristvey, A.G., Ross, D.S. & Kantor. G.F. 2009. Deployment of Wireless Sensor Networks for Irrigation and Nutrient Management in Nursery and Greenhouse Operations. *Proc. Southern Nursery Assoc. Res. Conf.* 54: 28-34.

Lea-Cox, J.D., Kantor, G.F., Bauerle, W.L., van Iersel, M.W., Campbell, C., Bauerle, T.L., Ross, D.S., Ristvey, A.G., Parker, D., King, D., Bauer, R., Cohan, S. M., Thomas, P. Ruter, J.M., Chappell, M., Lefsky, M., Kampf, S. & L. Bissey. 2010a. A Specialty Crops Research Project: Using Wireless Sensor Networks and Crop Modeling for Precision Irrigation and Nutrient Management in Nursery, Greenhouse and Green Roof Systems. *Proc. Southern Nursery Assoc. Res. Conf. 55*: 211-215.

Lea-Cox, J.D., Kantor, G.F., Bauerle, W.L., van Iersel, M.W., Campbell, C., Bauerle, T.L., Parker, D., King, D., Bauer, R., L. Bissey & Martin, T. 2010b. SCRI-MINDS-Precision Irrigation and Nutrient Management in Nursery, Greenhouse and Green Roof Systems: Year 1 Report. Available from <http://www.smart-farms.net/impacts> Accessed 08/02/2011.

Lea-Cox, J.D., Arguedas-Rodriguez, F. R.,  Ristvey A. G. & Ross. D. S. 2011. Relating Real-time Substrate Matric Potential Measurements to Plant Water Use, for Precision Irrigation. *Acta Hort.* 891:201-208.

Leith, J.H. & Burger. D.W. 1989. Growth Of Chrysanthemum Using An Irrigation System Controlled By Soil Moisture Tension. *J. Amer. Soc. Hort. Sci.* 114: 387-397

Liu, G. & Ying, Y. 2003. Application Of Bluetooth Technology In Greenhouse Environment, Monitor And Control. *J. Zhejiang Univ., Agric Life Sci.* 29:329-334.

Majsztrik, J.C., A.G. Ristvey, and J.D. Lea-Cox. 2011. Water and Nutrient Management in the Production of Container-Grown Ornamentals. *Hort. Rev.* 38:253-296.

Murray, J.D., Lea-Cox J.D. & Ross. D.S. 2004. Time Domain Reflectometry Accurately Monitors and Controls Irrigation Water Applications *Acta Hort.* 633:75-82.

Nemali, K.S. & van Iersel, M.W. 2006. An Automated System for Controlling Drought Stress and Irrigation in Potted Plants. *Sci. Hortic.* 110:292-297.

Nobel, P.S. 2009. Physicochemical and Environmental Plant Physiology 4th Ed. *Elsevier Inc.* Waltham, MA.

O'Shaughnessy, S.A. & Evett, S.R. 2008. Integration of Wireless Sensor Networks into Moving Irrigation Systems for Automatic Irrigation Scheduling. *Amer. Soc. Agric. Biol. Eng.* 29 June - 2 July, 2008. Providence, RI. Paper # 083452, 21p.

Pieruschka, R., Huber, G. & Berry, J.A. 2010. Control of Transpiration by Radiation. *Proc. Natl. Acad. Sci.* 107:13372-13377.

Raviv, M., Lieth, J.H. & Wallach, R. 2000. Effect of Root-Zone Physical Properties of Coir and UC Mix on Performance of Cut Rose (cv. Kardinal). *Acta Hort.* 554:231-238.

Ristvey, A.G., Lea-Cox, J.D. & Ross, D.S. 2004. Nutrient Uptake, Partitioning And Leaching Losses From Container-Nursery Production Systems. *Acta Hort.* 630:321-328.

Ross, D. S., Lea-Cox, J.D. & Teffeau, K.M. 2001. The Importance of Water in the Nutrient Management Process. *Proc. Southern Nursery Assoc. Res. Conf.* 46:588-591.

Ruiz-Garcia, L., Lunadei, L., Barreiro, P. & Robla, J. I. 2009. A Review of Wireless Sensor Technologies and Applications in Agriculture and Food Industry: State of the Art and Current Trends. *Sensors* 9: 4728-4750.

Secchi, S., Gassman, P.W., Jha, M., Kurkalova, L., Feng, H.H., Campbell, T. & King. C.L. 2007. The Cost of Cleaner Water: Assessing Agricultural Pollution Reduction at The Watershed Scale. *J. Soil Water Conservation* 62:10–22.

Smith, K.A. & Mullins, C. 2001. Soil and Environmental Analysis: Physical Methods. 2nd Ed. *Marcel Decker*, New York, NY.

Smajstrla A.G. & Harrison, D.S. 1998. Tensiometers for Soil Moisture Measurement and Irrigation Scheduling. *Univ. Fl. IFAS Ext. Cir.* No. 487. 8p.

Stanghellini, C. & van Meurs., W.T.M., 1989. Crop Transpiration: A Greenhouse Climate Control Parameter. *Acta Hort.,* 245: 384-388.

Stanghellini, C., Pardossi, A. & Sigrimis, N. 2007. What Limits the Application of Wastewater and/or Closed Cycle in Horticulture? *Acta Hort.* 747:323-330.

Starry, O., Lea-Cox, J.D., Ristvey, A.G. & Cohan, S. 2011. Utilizing Sensor Networks to Assess Stormwater Retention by Greenroofs. *Amer. Soc. Agric. Biol. Eng.* 7-12th August, 2011. Louisville, KY. Paper #1111202, 7p.

Taiz, L. & Zeiger, E. 2006. Plant Physiology. *Sinauer Associates, Inc.* Sunderland, MA

Testezlaf, R., Larsen, C.A., Yeager, T.H. & Zazueta. F.S. 1999. Tensiometric Monitoring of Container Substrate Moisture Status. *HortTechnology* 9:105-109.

Treder, J., Matysiak,B., Nowak, J. & Treder, W. 1997. Evapotranspiration and Potted Plants Water Requirements as Affected by Environmental Factors. *Acta Hort.* 449:235-240.

Tolle, G., Polastre, J., Szewczyk, R., Culler, D., Turner, N., Tu, K., Burgess, S., Dawson, T., Buonadonna, P. Gay, D. & Hong. W. 2005. A Macroscope in the Redwoods. *In: Proc. 3rd ACM International Conference on Embedded Networked Sensor Systems* (SenSys), 2005. San Diego. Nov 2-4, 2005. pp. 51- 63.

Topp, G.C. 1985. Time-Domain Reflectometry (TDR) and its Application to Irrigation Scheduling. *Adv. Irr.* 3:107-127.

Tyler, H.H., Warren, S.L & Bilderback, T.E 1996. Cyclic Irrigation Increases Irrigation Application Efficiency and Decreases Ammonium Losses. *J. Environ. Hort.* 14:194–198.

United States Global Change Research Program. 2011. Overview: Water. Available from <http://www.globalchange.gov/component/content/article/52-reports-and-assessments/481-overview-water> Accessed 07/15/2011.

van Iersel, M.W. 2003. Short-Term Temperature Change Affects the Carbon Exchange Characteristics and Growth of Four Bedding Plant Species. *J. Amer. Soc. Hort. Sci.* 128:100-106.

van Iersel, M. W., Seymour, R.M., Chappell, M., Watson, F. & Dove, S. 2009. Soil Moisture Sensor-Based Irrigation Reduces Water Use and Nutrient Leaching in a Commercial Nursery. *Proc. Southern Nursery Assoc. Res. Conf.* 54:17-21.

van Iersel, M.W., Dove, S. Kang, J-G. & Burnett, S.E. 2010. Growth and Water Use of Petunia as Affected by Substrate Water Content and Daily Light Integral. *HortScience* 45:277-282.

van Iersel, M.W., Dove, S. & Burnett, S.E. 2011. The Use of Soil Moisture Probes for Improved Uniformity and Irrigation Control in Greenhouses. *Acta Hort.* 893:1049-1056.

Vellidis, G., Tucker, M., Perry, C., Wen, C. & Bednarz, C. 2008. A Real-Time Wireless Smart Sensor Array for Scheduling Irrigation. *Comput. Electron. Agric.* 61:44-50.

Wang, Y.P. & Jarvis. P.G. 1990. Description and Validation of an Array Model-MAESTRO. *Agricultural Forest Meteorology*. 51:257–280.

Wang, C., Zhao, C.J., Qiao, X.J., Zhang, X. & Zhang, Y.H. 2008. The Design of Wireless Sensor Networks Node for Measuring the Greenhouse's Environment Parameters. *Computing. Technol. Agric. 259*:1037-1046.

Warsaw, A. L., Fernandez, R. T., Cregg, B. M. & Andresen. J. A. 2009. Water Conservation, Growth, and Water Use Efficiency of Container-Grown Woody Ornamentals Irrigated Based on Daily Water Use. *HortScience* 44: 1308-1318.

Yeager, T.H., Bilderback, T.E., Fare, D., Gilliam, C., Lea-Cox, J.D., Niemiera, A.X., Ruter, J.M., Tilt, K., Warren, S.L., Whitwell T. & Wright. R.D. 2007. Best Management Practices: Guide for Producing Nursery Crops. 2nd Ed. *Southern Nursery Assoc.*, Atlanta, GA.

Yoo, S., Kim, J., Kim, T., Ahn, S., Sung, J. & Kim, D. 2007. A2S: Automated Agriculture System Based on WSN. *In: ISCE 2007. IEEE International Symposium on Consumer Electronics*. Irving, TX, USA.

Zazueta, F.S., Yeager, T., Valiente, J.I. & Brealey, J.A. 1994. A Modified Tensiometer for Irrigation Control in Potted Ornamental Production. *Proc. Soil Crop Sci. Soc. Fla.* 53:36-39.

# 10

# A Low Cost Remote Monitoring Method for Determining Farmer Irrigation Practices and Water Use

Kristoph-Dietrich Kinzli
*Florida Gulf Coast University*
*USA*

## 1. Introduction

Irrigated agriculture has traditionally been the backbone of the rural economy and provides nearly 70% of the worlds food supply using only 30% of planted agricultural land. Irrigated agriculture in general is a large water user that consumes roughly 80% of freshwater supplies worldwide and in the Western United States (Oad et al. 2009; Oad and Kullman, 2006). Since irrigated agriculture uses a large and visible portion of surface water in the world and the Western United States, it is often targeted for increased efficiency to free water for other uses. Due to fish and wildlife concerns, and demands from a growing urban population, the pressure to reduce consumption by irrigated agriculture increases every year. As the world population continues to grow, irrigated agriculture will also need to meet the additional food production required. The current belief is that irrigated agriculture will need to maximize the crop per drop to meet the demand in the future. The problem lies in the fact that available water supplies are currently developed and new untapped sources are limited. In order to increase production with the current amount of available water and deal with external pressure for reduced water usage, irrigated agriculture has to become more efficient in its on-farm water application and its deliveries on a whole system scale. Decision Support Systems and modernization of infrastructure can be used on a large system scale to increase the efficiency of water deliveries and have been utilized successfully in New Mexico, China, Spain and Argentina (Oad et al. 2009; Gensler et al. 2009; FAO, 2006; Gao, 1999; FAO, 1994). The problem with large infrastructure improvement projects and large scale implementation of decision support systems is that significant capital is required in addition to organizational structures that allow for such massive undertakings.

One sector where irrigated agriculture can significantly reduce water usage and stretch every drop is in on-farm water delivery. Achieving high water use efficiencies on farm requires detailed knowledge about soil moisture and water application rates to optimally manage irrigation. The problem with achieving improved efficiency is that high efficiency is generally coupled with high cost on-farm monitoring systems. Such high cost monitoring set ups are generally prohibitive to small farmers in the United States and in irrigated areas throughout the world. Additionally, the traditional methods of measuring water application require a constant presence on farm and do not allow for remote monitoring.

This chapter will focus on a low cost methodology utilized to remotely instrument eight farm fields in the Middle Rio Grande Valley. The chapter will describe in detail the instrumentation of the farm fields including soil moisture sensors and low cost flow measuring devices. The chapter will also present results regarding water usage and farmer irrigation practices that were obtained from the low cost instrumentation method. It is the hope of the author that this type of low cost monitoring network finds acceptance and contributes to improvements in water use efficiency throughout the American West and beyond, allowing irrigated agriculture to meet growing demand in the future with limited water supplies.

## 2. Background

The Middle Rio Grande Conservancy District (MRGCD) may be one of the oldest operating irrigation systems in North America (Gensler et al. 2009). Prior to Spanish settlement in the 1600s the area was being flood irrigated by the native Pueblo Indians. At the time of Albuquerque's founding in 1706 the ditches, that now constitute the MRGCD, were already in existence and were operating as independent acequia (tertiary canal) associations (Gensler et al. 2009). In 2010 the MRGCD operated and maintained nearly 1,500 miles of canals and drains throughout the valley in addition to nearly 200 miles of levees for flood protection. The MRGCD services irrigators from Cochiti Reservoir to the Bosque del Apache National Wildlife Refuge. An overview map of the MRGCD is displayed in Figure 1. Irrigation structures managed by the MRGCD divert water from the Rio Grande to service agricultural lands, that include both small urban landscapes and large scale production of alfalfa, corn, vegetable crops such as chili and grass pasture. The majority of the planted acreage, approximately 85%, consists of alfalfa, grass hay, and corn which can be characterized as low value crops. In the period from 1991 to 1998, USBR crop production and water utilization data indicate that the average irrigated acreage in the MRGCD, excluding pueblo lands, was 53,400 acres (21,600 ha) (Kinzli 2010). Analysis from 2003 through 2009 indicates that roughly 50,000 acres (20,200 ha) are irrigated as non-pueblo or privately owned lands and 10,000 acres (4,000 ha) are irrigated within the six Indian Pueblos (Cochiti, San Felipe, Santo Domingo, Santa Ana, Sandia, and Isleta). Agriculture in the MRGCD is a $142 million a year industry (MRGCD, 2007). Water users in the MRGCD include large farmers, community ditch associations, six Native American pueblos, independent acequia communities and urban landscape irrigators. The MRGCD supplies water to its four divisions -- Cochiti, Albuquerque, Belen and Socorro -- through Cochiti Dam and Angostura, Isleta and San Acacia diversion weirs, respectively (Oad et al. 2009; Oad et al. 2006; Oad and Kinzli, 2006). In addition to diversions, all divisions except Cochiti receive return flow from upstream divisions.

Return flows are conveyed through interior and riverside drains. From the drains, excess water is diverted into main canals in the downstream divisions for reuse or eventual return to the Rio Grande. Drains were originally designed to collect excess irrigation water and drain agricultural lands, but are currently used as interceptors of return flow and as water conveyance canals that allow for interdivisional supply.

Water in the MRGCD is delivered in hierarchical fashion; first, it is diverted from the river into a main canal, then to a secondary canal or lateral, and eventually to an acequia or small ditch. Figure 2 displays the organization of water delivery in the MRGCD. Conveyance canals in the MRGCD are primarily earthen canals but concrete lined canals exist in areas where bank stability and seepage are of special concern. After water is conveyed through laterals it is delivered to the farm turnouts with the aid of check structures in the lateral canals. Once water passes the farm turnout it is the responsibility of individual farmers to apply water and it is applied to fields using basin or furrow irrigation techniques. The overall average yearly water diversion by the MRGCD is approximately 350,000 Acre-feet (Kinzli 2010).

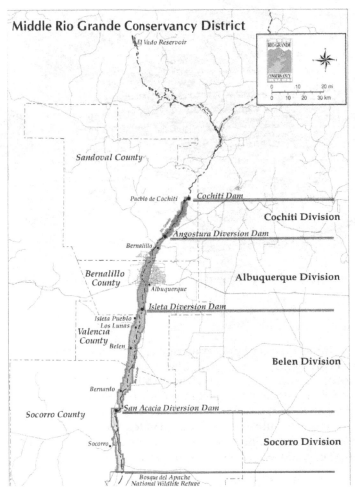

Fig. 1. Overview Map of MRGCD (MRGCD, 2007)

The MRGCD like many other conservancy districts has come under pressure to become a more efficient water user. In order to do so large scale infrastructure modernization projects have been undertaken (Gensler et al. 2009) and a decision support system has been developed and implemented (Kinzli, 2010). The one sector remaining where water saving can be realized is at the farm level by improving farmer irrigation application efficiency.

Measurements of on-farm application efficiency in the MRGCD were limited and therefore in the summer of 2008 eight fields in the MRGCD were instrumented to measure total water application and application efficiency. Figure 3 displays a map of the eight instrumented fields.

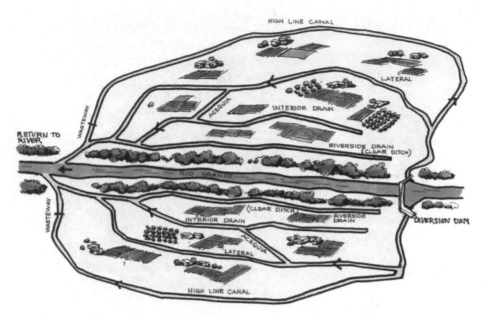

Fig. 2. Representation of MRGCD Irrigation System (Courtesy of David Gensler and MRGCD)

## 3. Methodology

In order to measure total water application and application efficiency it was necessary to instrument the eight fields with both a flow measurement device and instruments to measure soil moisture. Since the main crops in the MRGCD are alfalfa and grass hay 4 fields of each were chosen for monitoring. Due to the financial constraint of limited funding it was necessary to utilize a low cost setup with the total cost for each field remaining under $1200. This financial constraint would be a realistic consideration for farmers in the MRGCD as well since they produce low value crops such as alfalfa and grass hay. The use of a low cost monitoring network would also allow for application worldwide, specifically in developing countries.

### 3.1 Flow measurement

The first step in the field instrumentation was to perform a survey to determine the slope of the irrigation head-ditch, which was conducted using a laser level. The irrigation head ditches in the MRGCD are trapezoidal and have a 1 foot bottom width and a 1:1 H:V side slope. In addition to the survey, the dimensions of each head ditch were also determined. During the first irrigation event, the flow rate used for irrigation was measured using a Price

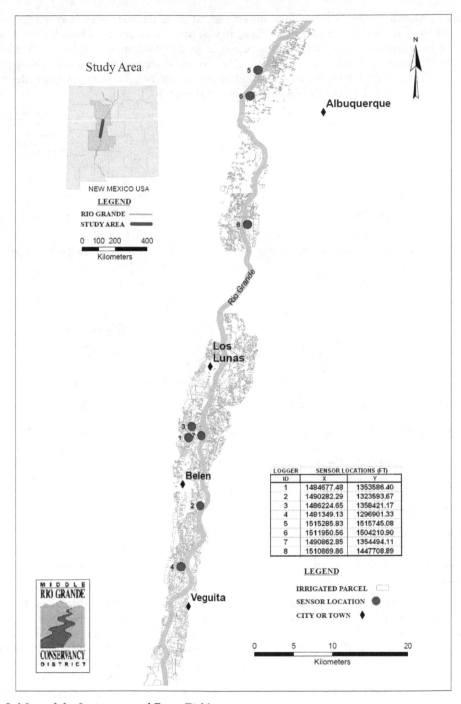

Fig. 3. Map of the Instrumented Farm Fields

Pygmy or Marsh McBirney flow meter and standard USGS measuring techniques. From the collected flow measurement and ditch data, it was possible to design a broad crested weir for flow measurement using the Unites States Bureau of Reclamation software Winflume and the Manning's flow rate equation. The software allows the user to design the appropriate flume and develops a stage-discharge equation based on the head over the crest of the weir. Figure 4 displays the flume designed for Field 3.

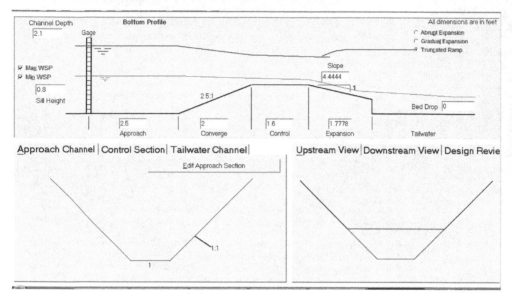

Fig. 4. Flume Designed for Measuring Field 3 (WSP = Water Surface Profile)

The broad crested weirs for each field were constructed out of concrete using cutout particle board templates as forms and cost approximately $100 each. Broad crested weirs were constructed for each of the eight fields but were utilized on seven fields. One farmer complained that the weir diminished his available flow rate, and therefore a rating curve was developed for his canal section instead of the weir. Figure 5 displays the finished broad crested weir for Field 3.

Hobo pressure transducers and data loggers ($400), manufactured by Onset Incorporated, were installed to measure the depth of water over the crest of the weir. These pressure transducers have an accuracy of 0.01 ft. Figure 6 displays a HOBO pressure transducer.

The Hobo dataloggers were installed on the side of each irrigation ditch roughly two canal widths upstream of the weir crest (Winflume design standard) using a small length of PVC pipe, clips, and concrete anchors. The section of PVC pipe was perforated multiple times with a ¼ inch drill bit to insure that water would seep into the section of PVC and allow for pressure measurement. Once the Hobo data loggers were installed in the PVC pipe a laser level was used to determine the offset between the bottom of the pressure transducer and the top of the weir crest. Figure 7 displays an installed Hobo pressure transducer.

Fig. 5. Completed Broad Crested Weir for Field 3

Fig. 6. Hobo Pressure Transducer

Fig. 7. Installed Hobo Pressure Transducer

The Hobo data loggers were set to log the absolute pressure every ten minutes. During an irrigation event, the pressure read by the Hobo included atmospheric pressure, so the atmospheric pressure from a Hobo exposed to only the atmosphere was subtracted from the reading. This resulted in a pressure reading that represented the total depth of water in the irrigation ditch. The pressure reading was converted using the conversion factor that 1 psi is the equivalent of 27.68 inches of water. Once the total depth of water in the irrigation ditch was calculated, the previously mentioned laser leveled offset of the weir crest was subtracted from the total water depth to get the depth of water over the weir crest. This value was plugged into the weir flow rate equation developed in Winflume to determine the flow passing the weir every ten minutes. Once the first irrigation event had occurred for each broad crested weir, the flow measurements calculated using the equation were compared to the initial measurement using flow meters in order to insure that the weirs were functioning properly. For each constructed weir the flow rate given by the Winflume equation was reasonable and corresponded to the measurements obtained using flow meters. The nature of this setup allowed for remote monitoring in that no on-farm presence was required during any irrigation event during the irrigation season. The Hobo pressure transducer has the capability to store an entire years worth of irrigation data and therefore data was only collected infrequently. When data was collected the Hobo optical USB cable was utilized to connect the pressure transducer to a laptop.

The total water volume in cubic feet applied during each irrigation event was obtained incrementally for every ten minute period during the irrigation event. The total volume in cubic feet was calculated by taking the flow rate in cubic feet per second every ten minutes and multiplying this value by 600 seconds. This was done for every ten minute interval during the duration of the irrigation event to obtain the total cubic feet of water applied during the event. This assumption to use a ten minute interval was validated by the fact that the water level did not fluctuate significantly during most irrigation events.

## 3.2 Soil moisture measurement

To improve irrigation efficiency the amount of moisture that is stored in the soil for beneficial plant use during the irrigation event and the subsequent depletion of the moisture is required. To measure the soil moisture, soil moisture probes were installed in each of the eight fields. During early 2008 before the irrigation season, soil moisture probes were installed in the eight representative fields instrumented with broad crested weirs. Electrical conductivity sensors were used instead of time domain reflectrometry (TDR) sensors due to budget constraints. TDR sensors can cost over $2000 a piece greatly, exceeding the budget available for each field. The electrical conductivity sensors used were the EC-20 ECHO probe from Decagon ($100 each). Figure 8 displays the EC-20 soil moisture probe.

Fig. 8. EC-20 ECHO Probe from Decagon Devices

Recent improvements to the ECH20 soil moisture sensor allowed for detailed measurement of soil water content (Sakaki et al. 2008). The ECH2O EC-20, which offers a low cost alternative to other capacitance type meters, (Kizito et al. 2008; Saito et al. 2008; Sakaki et al. 2008; Bandaranayake et al. 2007; Nemali et al. 2007; Plauborg et al. 2005) has been used to improve irrigation management for citrus plantations (Borhan et al. 2004). The precision of the ECH20 EC-20 is such that it can be used for greenhouse operations and to schedule field irrigation (Nemali et al. 2007). The main benefit of the ECH2O sensor is that it is one of the most inexpensive probes available and therefore can be widely used and implemented (Christensen, 2005; Luedeling et al. 2005; Riley et al. 2006). The ECH2O sensor is designed to be buried in the soil for extended periods of time and connected to a data logger such as the Em5b (Decagon Devices, Pullman WA). EC-20 sensors allow for the determination of saturation, field capacity, and wilting point, along with the redistribution pattern of soil water, and possible drainage below the root zone. This information can be used to decide the time and amount of irrigation (Bandaranayake et al. 2007).

The EC-20 probe has a flat design for single insertion and allows for continued monitoring at a user defined interval. The overall length of the sensor is 8 inches with a width of 1.2 inches and blade thickness of 0.04 inches, with a 2.4 inch sensor head length. The total sampling volume of the probe is between 7.8 and 15.6 in$^3$, depending on soil water content (Bandaranayake et al. 2007). The ECH2O EC-20 soil probe measures the dielectric permittivity or capacitance of the surrounding soil medium, and the final output from the sensor is either in a millivolt or raw count value that can be converted to a volumetric water content using calibration equations (Kelleners et al. 2005). The raw count is an electrical output specific to which datalogger the sensor is used with. Raw counts can easily be converted if an output in millivolts is desired. Details on the EC-20 sensor measurement principle and function are reported by the manufacturer (Decagon Devices, 2006a). Studies have shown that temperature affects on the ECH2O probes are minimal (Kizito et al. 2008; Norikane et al. 2005; Campbell, 2002) with changes of 0.0022 ft$^3$/ft$^3$ water content per degree C (Nemali et al. 2007). Problems due to soil variation and air gaps can be avoided by using the factory installation tool and developing calibration equations relevant to each soil type. Drawbacks of this sensor include water leakage into the sensor circuit in isolated cases, and damage from animals such as gophers and squirrels (Bandaranayake et al. 2007). Using the manufacturer provided equation, typical accuracy in medium textured soil is expected to be ±0.04 ft$^3$/ft$^3$ (3% average error) with soil specific equations producing results with an accuracy of ±0.02 ft$^3$/ft$^3$ (1% average error) (Decagon Devices, 2006b).

Through previous research it has been found that dielectric sensors often require site specific calibration either through field methods or laboratory analyses. Inoue et al (2008) and Topp et al (2000) found that it was necessary to perform site specific calibrations for capacitance sensors to account for salinity concerns, and Nemali et al (2007) found that it was necessary to calibrate the ECH2O sensors because output was significantly affected by the electrical conductivity of the soil. Other studies have found that site specific corrections are required for mineral, organic, and volcanic soils (Paige and Keefer, 2008; Bartoli et al. 2007; Regelado et al. 2007; Malicki et al. 1996).

Kizito et al. (2008) suggested that soil specific calibrations are important when large networks of ECH2O soil moisture sensors are deployed. Several researchers have found that soil specific calibrations are necessary for ECH20 probes across varying soil types (Sakaki et al. 2008; Mitsuishi and Mizoguchi, 2007; Fares and Polyakov, 2006; Bosch, 2004) and Saito et al (2008) found that calibration is a requirement for accurate determination of volumetric water content using the ECH2O. Based on the recommendations of these previous studies, soil specific calibrations were performed for each sensor installation using a technique described in (Kinzli, 2010). The use of EC20 ECHO sensors allowed for development of a low cost monitoring network capable of being used to schedule irrigation and therefore offer the possibility of improving water use efficiency.

The EC-20 ECHO probes installed in the eight fields were linked to Em5b data recorders ($400 each). The Em5b is a 5-channel, self-contained data recorder (Decagon, 2008). The Em5b is housed in a white UV-proof enclosure, which makes it suitable for general outdoor measurements. It uses 4 AAA-size alkaline batteries, that last 5-6 months, and has a Flash Data memory that allows for 145 days of data collection at 1 scan/hour (Decagon, 2008). All eight Em5b data loggers were set to record soil moisture every 60 minutes during the study. Figure 9 displays the Em5b data logger.

Fig. 9. Em5b-Datalogger from Decagon Devices

The EC-20 ECHO moisture probes were installed in the eight representative fields to obtain a value of soil moisture remaining before an irrigation event and to determine hourly soil moisture depletions. Each field was equipped with one sensor station, due to project budget constraints. Therefore each field represented a point measurement. This approach resulted in eight point measurements throughout the MRGCD. Lundahl (2006) showed that soil moisture measurements at one point in each field were sufficient to obtain soil moisture depletion and application efficiency in the MRGCD. The field layout used for each sensor station is displayed in Figure 10. The layout of the moisture probes was designed to eliminate data points in areas that display variable wetting front values due to distance and the points chosen provided average values for the field in question. Each sensor station consisted of two EC-20 ECHO probes (installed at 8 inches and 24 inches) so that a soil profile of up to 4 feet could be measured. Figure 11 displays the layout of a sensor station.

The Em5b data loggers were located outside of the field boundary to minimize interference with cultivation and prevent damage of the logger. A 50 ft extension cable was used to place the sensor stations out in the field to eliminate edge effects on crop ET. The 50 ft extension cable was placed in a hand dug trench out into the field at a depth of roughly 8 inches.

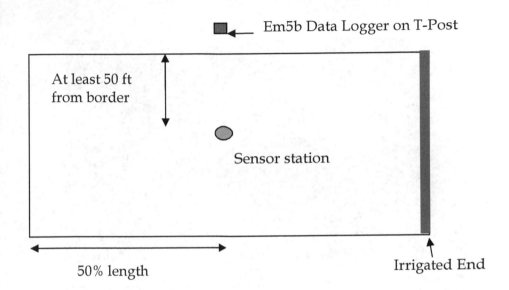

Fig. 10. Field Layout of Sensor Stations

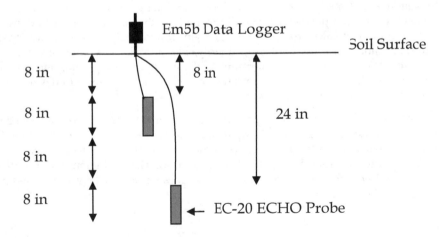

Fig. 11. Individual Layout of Sensor Stations

Once the soil moisture probes were installed, GPS points were taken at the location of the sensor station, datalogger, and the corner of the field to determine the exact irrigated acreage. From this collected data and the MRGCD aerial photography coverage, a detailed map of the flume, sensor, and datalogger location was created for each of the eight study fields. Figure 12 displays the map of Field 3.

LOGGER ID - 3

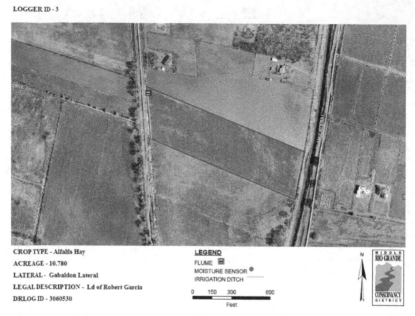

CROP TYPE - Alfalfa Hay
ACREAGE - 10.780
LATERAL - Gabaldon Lateral
LEGAL DESCRIPTION - Ld of Robert Garcia
DRLOG ID - 3060530

LEGEND
FLUME
MOISTURE SENSOR
IRRIGATION DITCH

0    150   300        600
Feet

N

MIDDLE
RIO GRANDE

CONSERVANCY
DISTRICT

Fig. 12. Map of Study Field 3

In order to validate that the probes were indeed functioning correctly and to develop calibration equations soil samples were taken in proximity to the installed sensor stations. A one gallon soil sample was taken for each installed sensor and analyzed at Colorado State University to determine soil type, field bulk density, pH, and electrical conductivity. Soil samples were also taken in order to determine field capacity, wilting point, and readily available moisture (RAM) for each soil type. These samples were also used to develop soil specific calibration equations for each sensor.

Using the instrumented fields it was possible to determine the on-farm application efficiency over a period of 2 years and 144 irrigation events for the eight instrumented fields. For the purpose of this analysis the on farm application efficiency was defined as the water replenished for crop use divided by the total water applied. This definition of application efficiency focuses only on water for crop growth   and does not include any water used for leaching salts out of the root zone.

## 4. Results

In order to determine application efficiency the broad crested weirs and pressure transducers installed on the eight farm fields were used to determine the total water delivered for each irrigation event during the 2008 and 2009 irrigation seasons. Once the total water applied for an irrigation event was calculated, it was possible to calculate the depth of water applied per unit area by dividing the total volume applied by the acreage of the basin that was irrigated. This resulted in a depth of water in inches applied over the monitored field. Additionally, irrigation event number, the date, duration, and average flow rate for each irrigation event were recorded. Table 1 displays the logger ID, irrigation event, irrigation date, total water applied, and inches applied for ten irrigation events.

| Logger ID | Irrigation Event | Date | Total Water Applied (ft³) | Depth Applied (inches) |
|-----------|------------------|------|---------------------------|------------------------|
| 1 | 1 | 4/14/2008 | 157190 | 6.95 |
| 1 | 2 | 5/5/2008 | 266004 | 7.44 |
| 1 | 3 | 6/1/2008 | 325216 | 9.09 |
| 1 | 4 | 6/24/2008 | 149748 | 4.19 |
| 1 | 5 | 8/6/2008 | 150338 | 4.2 |
| 1 | 6 | 9/12/2008 | 125121 | 3.5 |
| 1 | 1 | 4/13/2009 | 112475 | 3.15 |
| 1 | 2 | 5/11/2009 | 148812 | 4.16 |
| 1 | 3 | 6/18/2009 | 173791 | 4.86 |
| 1 | 4 | 7/20/2009 | 113443 | 3.17 |

Table 1. Logger ID, Irrigation Event, Date, Total Water Applied and Depth Applied for 10 Irrigation Events

The next step in calculating the application efficiency was determining the water available for crop use that was replenished during each irrigation event. This was possible using the data collected from the installed EC-20 soil moisture sensors. The soil moisture sensor data, corrected using the developed laboratory calibration equations for each specific sensor installation, provided the volumetric soil moisture content before the irrigation event and after field capacity was reached. The difference between the volumetric water content before the irrigation event and field capacity represented the amount of water stored in the root zone for beneficial crop use. This data was recorded at both the 8 inch and 24 inch sensor location for each field for each irrigation event. To calculate the water stored in the soil for beneficial crop use in inches the 8 inch sensor was deemed to be representative of the first 16 inches of root depth for both the alfalfa and grass hay fields. The 24 inch sensor was chosen to represent the subsequent 20 inches of root depth for grass hay and the subsequent 32 inches for alfalfa. For grass hay and alfalfa this represented a 36 inch and 48 inch effective total root zone, respectively. These values were chosen based on 12 years of research conducted by Garcia et al. (2008) at the Natural Resource Conservation Service (NRCS), which was conducted in the Middle Rio Grande and Mesilla Valleys to determine the root depths that were effectively able to utilize and deplete soil moisture.

Once the effective root depth was determined, the root depth associated with each sensor and crop type was multiplied by the difference between the volumetric water content at field capacity and volumetric water content before the irrigation event took place for the 8 inch and 24 inch sensor. This yielded the water available for crop use in inches for the upper

16 inches and either lower 20 inches for grass hay or 32 inches for alfalfa. These two values were added together to give the total water in inches available for crop use applied during the irrigation event. The total water available for crop use was then divided by the total water applied to determine application efficiency. The application efficiency for all 144 irrigation events was calculated from the collected data. **Table 2** displays the results of the application efficiency analysis for 10 irrigation events.

| Logger ID | Irrigation Event | Date | Depth Applied (inches) | Moisture Applied for Crop Use (inches) | Application Efficiency (%) |
|-----------|------------------|------|------------------------|----------------------------------------|----------------------------|
| 1 | 1 | 4/14/2008 | 6.95 | 4.64 | 67% |
| 1 | 2 | 5/5/2008 | 7.44 | 1.92 | 26% |
| 1 | 3 | 6/1/2008 | 9.09 | 3.36 | 37% |
| 1 | 4 | 6/24/2008 | 4.19 | 2.24 | 53% |
| 1 | 5 | 8/6/2008 | 4.2 | 1.76 | 42% |
| 1 | 6 | 9/12/2008 | 3.5 | 2.4 | 69% |
| 1 | 1 | 4/13/2009 | 3.15 | 2.56 | 81% |
| 1 | 2 | 5/11/2009 | 4.16 | 2.56 | 62% |
| 1 | 3 | 6/18/2009 | 4.86 | 2.56 | 53% |
| 1 | 4 | 7/20/2009 | 3.17 | 1.12 | 35% |

Table 2. Irrigation Event, Date, Depth Applied, Moisture Applied for Beneficial Crop Use and Application Efficiency for 10 Irrigation Events

The data displayed significant variability with a range in application efficiency from 8% to 100%. The mean value for all 144 irrigation events was found to be 44.4% with a standard deviation of 24.4%. The calculated mean value represent a lower application efficiency value than the 50% previously hypothesized by water managers.

To address the variability in the collected data a histogram of the collected data was created. Figure 13 displays the histogram of application efficiency.

The developed histogram displayed a nearly normal distribution about the mean value but was skewed slightly to the right due to 11 irrigation events with an application efficiency of 100%. From the developed histogram it became clear that the majority of irrigation events exhibited application efficiencies reflected by the calculated mean value.

Using the developed histogram it was also possible to calculate the probability that the application efficiency would fall within one standard deviation of the calculated mean. The probability that the application efficiency of an irrigation event would fall within one standard deviation was found to be 112 out of 144 irrigation events resulting in a probability value of 0.78. This indicates that 78% of the irrigation events were within one standard deviation of the calculated mean. Based on the analysis of the histogram and probability the revised value for application efficiency of 45% will be utilized by the MRGCD, which will allow for more precise representation of farmer practices. Several irrigation events exhibited an application efficiency of 100% and indicate possible under irrigation. Such results also point to possible measurement errors and residual moisture that is used by plants but not accounted for in calculations related to an irrigation event. One reason for possible errors could be due to the fact that only one sensor location was installed for each field due to budget constraints. Spatial variability in soil and topography that could not be measured due to a single sensor location could be the cause of uneven water distribution during the irrigation event. Differences in moisture uptake by plants due to spatial root variability could also be the cause this discrepancy.

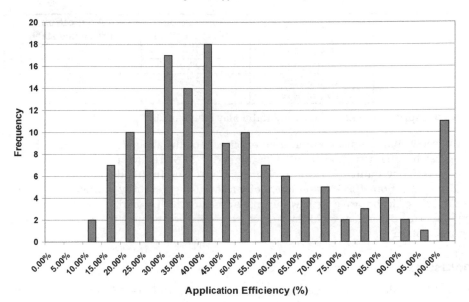

Fig. 13. Histogram of Application Efficiency

From the collected data it was also possible to refine the analysis of on farm application efficiency. First, the application efficiency was separated by crop type as analysis of the total water applied during an entire season suggested that fields with alfalfa hay would have higher application efficiency. The mean value of application efficiency for each grass field was calculated for the 2008 and 2009 irrigation seasons from all irrigation events. For 2008 the application efficiencies covered a range from 31% to 50%. For 2009 the application efficiency covered a range from 22% to 52%. The mean application efficiency of all 40 grass hay irrigation events was found to be 40.8% in 2008. The mean application efficiency of all 43 grass hay irrigation events was found to be 38.6% in 2009. Table 3 displays the average values found for each individual grass field.

The mean value of application efficiency for each alfalfa field was also calculated for the 2008 and 2009 irrigation seasons for all irrigation events. For 2008 the application efficiencies covered a range from 29% to 82%. For 2009 the application efficiency covered a range from 23% to 85%. The mean application efficiency of all 31 alfalfa hay irrigation events was found to be 50.2% in 2008. The mean application efficiency of all 30 alfalfa hay irrigation events was found to be 52.5% in 2009. Table 4 displays the average values found for each individual alfalfa field.

| Logger ID | Crop Type | Application Efficiency 2008 | Application Efficiency 2009 |
|-----------|-----------|----------------------------|----------------------------|
| 4 | Grass Hay | 50% | 52% |
| 5 | Grass Hay | 44% | 41% |
| 7 | Grass Hay | 33% | 36% |
| 8 | Grass Hay | 31% | 22% |

Table 3. Mean Application Efficiency for Grass Hay Fields in 2008 and 2009

| Logger ID | Crop Type | Application Efficiency 2008 | Application Efficiency 2009 |
|-----------|-----------|----------------------------|----------------------------|
| 1 | Alfalfa Hay | 49% | 66% |
| 2 | Alfalfa Hay | 29% | 23% |
| 3 | Alfalfa Hay | 82% | 85% |
| 6 | Alfalfa Hay | 45% | 43% |

Table 4. Mean Application Efficiency for Alfalfa Hay Fields in 2008 and 2009

The results show that the mean application efficiency for the alfalfa fields was 9.4% higher than the grass hay fields in 2008 and 13.9% higher in 2009. The temporal variation of the application efficiency numbers was also examined but no useful trends could be identified. Overall, the application efficiency numbers obtained during the study indicate that farmers in the MRGCD could improve their water management which would result in more water being available for other uses including increased production to meet the needs of our ever growing population.

## 5. Conclusions

As the world population continues to grow, irrigated agriculture will need to meet the additional food production required. The current belief is that irrigated agriculture will need to maximize the crop per drop to meet the demand in the future as current water supplies are already stretched thin. In order to increase production with the current amount of available water and deal with external pressure for reduced water usage, irrigated agriculture can become more efficient in its on-farm water application. Increasing on-farm application efficiency if often cost prohibitive, especially for low value crops. This chapter presented a low cost methodology utilized to remotely instrument eight farm fields in the Middle Rio Grande Valley for measurements of both applied irrigation water and soil moisture conditions. Through the instrumented fields it was possible to determine application efficiencies for 144 irrigation events over a period of 2 years. The total cost for each instrumented field was $1200 dollars and represents a cost level that most farmers in the Western United States could bear regardless of crop value.

The field instrumentation presented in this study was only used to monitor the eight fields and describe how farmers currently irrigate in the MRGCD. In order to achieve higher application efficiencies and obtain the most crop per drop the field instrumentation setup described in this chapter could be used to schedule irrigation events and precisely apply the appropriate amount of water. Knowledge of the soil moisture conditions prior to an irrigation event could be obtained from EC-20 sensor setups and an optimal application depth could be calculated. This application depth could then be precisely applied using the broad crested weirs placed in irrigation head ditches. It is the hope of the author that this type of low cost monitoring network finds acceptance and contributes to improvements in water use efficiency throughout the American West and beyond allowing irrigated agriculture to meet growing demand in the future with limited water supplies.

## 6. References

Bandaranayake, W.M., Borhan, M.S., and J.D. Holeton. 2007. Performance of a capacitance-type soil water probe in well-drained sandy soil. Soil Sci.Soc. Am. J. 71(3): 993 – 1002.

Bartoli, F., Regalado, C.M., Basile, A., Buurman, P. & Coppola, A. 2007. Physical properties in European volcanic soils: a synthesis and recent developments. In: Soils of volcanic regions in Europe, pp. 515–537. Springer-Verlag, Berlin.

Borhan, M.S., L.R. Parsons, and W. Bandaranayake. 2004. Evaluation of a low cost capacitance ECH2O soil moisture sensor for citrus in a sandy soil. p. 447–454. In Int. Irrig.

Bosch, D.D. 2004. Comparison of capacitance-based soil water probes in coastal plain soils. Vadose Zone Journal. 3: 1380-1389.

Campbell, C.S. 2002. Response of ECH2O Soil Moisture Sensor to Temperature Variation. Application Note. Decagon Devices, Pullman, WA.

Christensen, N.B., 2005. Irrigation management using soil moisture monitors. Western Nutrient Manage. Conf., 6: 46-53.

Decagon (2008). Environmental Research Instrumentation. ECH$_2$O Probe models EC-5, EC-10 and EC-20. http://www.decagon.com/echo/ec20.html

Decagon Devices, Inc., 2006a. ECH2O Dielectric Probes vs Time Domain Reflectometers (TDR). Decagon Devices Inc. Application Note. Decagon Devices, Inc., Pullman, Wash.

Decagon Devices, Inc., 2006b. ECH2O Soil Moisture Sensor Operator's Manual For Models EC-20, EC-10 and EC-5 Version 5. Decagon Devices, Inc., Pullman, WA. pp: 23.

Fares, A. and V. Polyakov, 2006. Advances in crop water management using capacitive water sensors. Adv. Agron. 90: 43-77.

Gao, Z. 1999. Decision making Support System for Irrigation Water management of Jingtai Chuan Pumping Irrigation Scheme at the Upper Reaches of Yellow River. Watsave Workshop Paper Presented at 51st IEC, Cape Town, South Africa

Garcia, R. 2008. New Mexico Integrated Water Management Handbook. United States Department of Agriculture, Natural Resources Conservation Service.

Gensler, D. Oad, R. and Kinzli, K-D. 2009. Irrigation System Modernization: A Case Study of the Middle Rio Grande Valley. ASCE Journal of Irrigation and Drainage. 135(2): 169-176.

Gensler, D. 2005. Interviews with David Gensler MRGCD Head Hydrologist, May-August

Inoue, M., Ould Ahmed, B.A., Saito, T., Irshad, M., and K.C. Uzoma. 2008. Comparison of three dielectric moisture sensors for measurement of water in saline sandy soils. Soil Use and Manag. 24: 156-162

Kelleners, T.J., D.A. Robins, P.J. Shouse, J.E. Ayars, and T.H. Skaggs. 2005. Frequency dependence of the complex permittivity and its impact on dielectric sensor calibration in soils. Soil Sci. Soc. Am. J. 69:67–76.

Kinzli, K-D. 2010. Improving Irrigation System Performance through Scheduled Water Delivery in the Middle Rio Grande Conservancy District. Dissertation. Colorado State University. May.

Kizito, F., Campbell, C.S., Campbell, G.S., Cobos, D.R., Teare, B.L., Carter, B., and J.W. Hopmans. 2008. Frequency, electrical conductivity and temperature analysis of a low-cost capacitance soil moisture sensor. J. of Hyd. 352 (3-4) 367-378.

Luedeling, E., M. Nagieb, F. Wichern, M. Brandt, M. Deurer and A. Buerkert, 2005. Drainage, salt leaching and physico-chemical properties of irrigated man-made terrace soils in a mountain oasis of northern Oman. Geoderma, 125: 273-285.

Lundahl, A. 2006. Quantifying, Monitoring, and Improving the Efficiency of Flood Irrigation in the Hydrosphere of Candelaria Farms Preserve, Albuquerque, New Mexico. Masters Thesis. Water Resources Program, University of New Mexico. Albuquerque, New Mexico.

Malicki, M.A., Plagge, R. & Roth, C.H. 1996. Improving the calibration of dielectric TDR soil moisture determination taking into account the solid soil. European Journal of Soil Science, 47, 357-366.

Mitsuishi, S., and M. Mizoguchi. 2007. Effects of Dry Bulk Density and Soil Type on Soil Moisutre Measurements using ECH2O-TE Probe. ASA-CSSA-SSSA International Annual Meeting. Nov 4-8. New Orleans, LA.

MRGCD. 2007. Middle Rio Grande Conservancy District – Keeping the Valley Green. Flyer of the Middle Rio Grande Conservancy District. Middle Rio Grande Conservancy District. Albuquerque, New Mexico.

Nemali, K.S, Montesano, F., Dove, S.K. and M. W. van Iersel. 2007. Calibration and performance of moisture sensors in soilless substrates: ECH2O and Theta probes. Sci. Hort. 112(2): 227-234

Norikane, J.H., J.J. Prenger, D.T. Rouzan-Wheeldon, and H.G. Levine. 2005. A comparison of soil moisture sensors for space fl ight applications. Appl. Eng. Agric. 21:211-216.

Oad, R. Garcia, L. Kinzli, K-D, Patterson, D and N. Shafike. 2009. Decision Support Systems for Efficient Irrigation in the Middle Rio Grande Valley. ASCE Journal of Irrigation and Drainage. 135(2): 177-185.

Oad, Ramchand and K. Kinzli. 2006. SCADA Employed in Middle Rio Grande Valley to Help Deliver Water Efficiently. Colorado Water – Neswletter of the Water Center of Colorado State University. April 2006,

Oad, Ramchand and R. Kullman. 2006. Managing Irrigated Agriculture for Better River Ecosystems—A Case Study of the Middle Rio Grande. Journal of Irrigation and Drainage Engineering, Volume 132, No. 6: 579-586.

Plauborg, F. 1995. Evaporation from bare soil in a temperature humid climate. Measurement using micro-lysimeters and TDR. Agriculture and Forest Meteorology, 76, 1-17.

Paige, G.B., and T.O. Keefer. 2008. Comparison of field performance of multiple soil moisture sensors in a semi-arid rangeland. J. Am. Water Resources Association. 44(1): 121-135.

Regalado, C.M., Ritter, A. & Rodrı́guez, G..R.M. 2007. Performance of the commercial WET capacitance sensor as compared with time domain reflectometry in volcanic soils. Vadose Zone Journal, 6, 244-254.

Riley, T.C., T.A. Endreny and J.D. Halfman, 2006. Monitoring soil moisture and water table height with a low-cost data logger. Comput. Geosci., 32: 135-140.

Saito, T., Fujimaki, H., and Inoue, Mitsuhiro. 2008. Calibration and Simultaneous Monitoring of Soil Water Content and Salinity with Capacitance and Four-electrode Probes. American Journal of Environmental Sciences 4 (6), 690 - 699

Sakaki, T., Limsuwat, A., Smits, K.M., and T.H. Illangasekare. 2008. Empirical two-point α-mixing model for calibrating the ECH2O EC-5 soil moisture sensor in sands. Water Resources Research 44: 1-8.

Topp, G.C., Zegelin, S. & White, I. 2000. Impact of real and imaginary components of relatively permittivity on time domain reflectometry measurement in soils. Soil Sci. Soc. Am. J. 64, 1244-1252.

# 11

# Critical Evaluation of Different Techniques for Determining Soil Water Content

Alejandro Zermeño-González[1], Juan Munguia-López[2],
Martín Cadena-Zapata[1], Santos Gabriel Campos-Magaña[1],
Luis Ibarra-Jiménez[2] and Raúl Rodríguez-García[1]

*[1]Antonio Narro Autonomous Agrarian University*
*[2]Department of Agricultural Plastics*
*Research Center for Applied Chemistry, Saltillo, Coahuila*
*México*

## 1. Introduction

To efficiently operate any type of irrigation system, it is necessary to know when to irrigate and the quantity of water to apply during irrigation. To achieve this, it is very important to know the previously available soil water content. A good on-farm irrigation water management requires a routine monitoring of soil water moisture. Soil water must be maintained between a lower and upper limit of availability for an optimum plant growth. Soil moisture is a very dynamic variable that depends on plants evapotranspiration, irrigation frequency, drainage and rainfall. Measuring soil water content for determining the water depth allows avoiding the economic losses due to the effect of underirrigation on crop yield and crop quality, and the environmentally costly effects of overirrigation on wasted water and energy, leaching of nutrients or agricultural chemicals into groundwater supplies.

This chapter describes the applications and limitations of different techniques for determining soil water moisture. A description of how to calculate the irrigation depth as a function of water soil holding capacity, soil depth and bulk density is also included. Six techniques for measurement of soil moisture are described: gravimetric sampling, neutron scattering, tensiometers, porous blocks, time domain reflectometry, impedance and capacitance methods.

## 2. State of water in the soil

The state of water in the soil can be described in two ways: quantity present and energy status. The quantity present is expressed as gravimetric (mass) or volumetric. The gravimetric water content is the mass of water in a unit mass of dry soil (g of water/g of dry soil). The wet weight of soil sample is determined; the sample is dried at 105 °C to constant weight and reweighed (Gardner, 1986). The volumetric water content is expressed in terms of the volume of water per volume of soil (cm$^3$ of water/cm$^3$ of soil). Volumetric water content can be calculated from gravimetric water using the equation:

$$\theta v = \theta w * \rho b \tag{1}$$

Where $\theta v$ is the volumetric water content, $\theta w$ is the gravimetric water content and $\rho b$ is the soil bulk density, which must be determined for the same soil under field conditions.

The energy status of water in soil can be expressed as follows (Hanks and Ashcroft, 1980):

$$\Psi_{total} = \Psi_{matric} + \Psi_{solute} + \Psi_{grav.} \tag{2}$$

Where $\Psi_{total}$ is the total soil water potential (MPa), $\Psi_{matric}$ soil matric potential (MPa), $\Psi_{solute}$ soil solute potential (MPa) and $\Psi_{grav}$ pressure potential or gravimetric water potential (MPa). The energy of water in the soil is attenuated by the hydrophilic surfaces of soil particles. As a result of the attraction of water to these surfaces, the energy of the water is decreased. Water forms films around the particles and fills pores. This fraction of the soil water energy is known as capillarity suction or matric water potential. The value of the matric term can be calculated from the capillarity rise equation:

$$\Psi_{matric} = -\rho_w g h = \frac{-2\gamma \cos(\alpha)}{r} \tag{3}$$

where, $\rho_w$ is the density of water (kg m$^{-3}$), h = height of rise above a free water surface (m), g = acceleration due to gravity ( m s$^{-2}$), $\gamma$ = surface tension (N m$^{-1}$), $\alpha$ = wetting angle (degrees) and r = capillarity radius (m)  (Hanks & Ashcroft, 1980; Hillel, 1980). The pressure potential is present in saturated soil due to the pressure of water above a given point and is calculated with the equation:

$$\Psi_{grav.} = \rho_w g h \tag{4}$$

where, $\rho_w$, g, and h were previously defined. The presence of solutes in the soil water further decreases its energy potential. The solute or osmotic potential of soil water is less than or equal to zero, and is directly related to the total solute concentration in the water, according to the following equation:

$$\Psi_{solute} = cRT \tag{5}$$

where, R is the universal gas constant (8.3143 J K$^{-1}$ mol$^{-1}$), T is the absolute temperature and c is the osmolality of the solution. At low concentrations, where the activity coefficient is near 1, c is approximately equal to the total molar concentration of osmotically active species in the water.

For a given soil, there is a unique relationship between the soil water content and the soil water potential. This relationship is known as the soil water characteristic curve or soil water release curve (Klute, 1986). The curve derived by determining the energy status of water in the soil at several water contents may vary considerably with changes in soil texture (Figure 1)

Two approaches are used to obtain the relationship between soil water content and soil water potential. Either a given water content is first established, and the water potential then determined, or conditions are imposed on a soil sample to bring it to a given water potential, and the water content of the sample is determined  after equilibrium is reached.

In the latter case, vacuum and pressure plate apparatus have been use extensively down to -1.5 MPa matric water potential (Klute, 1986). These are best applied where the soil solution is diluted and therefore the contribution of solutes to total water potential is minimal. In typical applications, moist soil samples are placed on a ceramic plate (down to -0.08 MPa) or a membrane (to -1.5 MPa), and a fixed suction or pressure is applied to a given potential until no more water is forced out of the sample. In practical terms, a vacuum can be applied to the ceramic plates down to potential of approximately -0.8 MPa. Below this potential, the soil samples must be housed in a pressure chamber to which constant air pressure can be applied; water is then force out of the soil sample and through the ceramic plate or membrane until no ore water is drained. At this point, it is assumed that the water potential of the remaining soil water is exactly equal to the negative of the pressure applied. This technique is used down to water potential of -1.5 MPa.

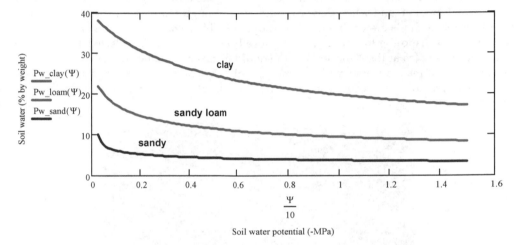

Fig. 1. Typical relationships between soil water content and soil water potential in clay, sandy loam and clay soils.

Psychrometric systems have been used to determine the total soil water potential of samples at different soil water contents. The relative humidity of air in equilibrium with the moist soil sample is determined, and expressed in terms of the corresponding water potential. If the soil is low in salts, only the matric potential is represented; otherwise, the sum of matric and osmotic potential results. Because relative humidity near 100% may be difficult to measure accurately, the psychrometric technique may be difficult to measure accurately, the psychrometric technique is best applied to systems where the soil water potential is less than -0.20 MPa. (Rundel & Jarrel, 1991)

## 2.1 Depth of available soil water

The Depth of total, depleted and residual available soil water can be calculated from the following equations:

$$TAW = [\theta_{w\_FC} - \theta_{w\_PWP}]*(\rho b/\rho w)*Z \tag{6}$$

$$DAW = [\theta_{w\_FC} - \theta_{w\_actual}]*(\rho b/\rho w)*Z \tag{7}$$

$$RAW = [\theta_{w\_actual} - \theta_{w\_PWP}]*(\rho b/\rho w)*Z \qquad (8)$$

where, TAW=depth of total available soil water (cm), DAW= depth of depleted available soil water (cm), RAW= depth of residual available soil water (cm), $\theta_{w\_FC}$ = gravimetric soil water content at field capacity (g/g), $\theta_{w\_PWP}$ = gravimetric soil water content at permanent wilting point (g/g), $\theta_{w\_actual}$ = gravimetric soil water content at the time of measurement (g/g), $\rho b$ = soil bulk density (g/cm³), $\rho w$ = density of water (g/cm³), Z = soil depth to irrigate (cm).

## 3. Gravimetric water content

Gravimetry refers to the measurement of soil water content by weighing. It is the oldest and most direct method, and when done carefully with enough samples is the standard against which other methods are calibrated and compared. This technique requires careful sample collection and handling to minimize water lose between the time is collected and weighed. Replicated samples at the same soil depth should be taken to reduce the inherent sampling variability that results from small volumes of soil. The equipment required includes a soil auger, sample collection cans, a balance accurate to at least 1 gram and a drying oven (Figure 2).

Fig. 2. Equipment used by the gravimetric technique for measuring soil water content.

The technique involves taking soil samples from each of several desired depths in the crop root zone and temporarily storing them in containers (water vapor-proof). The samples are then weighed and the opened containers oven-dried under specific time and temperature conditions (105 °C for 24 h). The dry samples are re-weighed. Percent soil water content on a dry mass or gravimetric basis, Pw is determined as:

$$Pw = \frac{WSW - DSW}{DSW} * 100 \qquad (9)$$

where, WSW = wet sample weight (g), DSW = dry sample weight (g). The difference between wet and dry weight is the mass of water remove by drying. To convert from gravimetric basis to water content on a volumetric basis (Pv), multiply the gravimetric soil water content by the soil bulk density (ρb).

$$Pv = Pw * \rho b \qquad (10)$$

Although the gravimetric method is relatively simple and inexpensive, it has several limitations. It is time-consuming and labor-intensive compared with other methods of soil moisture measurements, results are known after a minimum of 24 h after sampling, a large number of samples must be taken to remove the inherent variability of this approach. As it is a destructive technique, repeated measurements at the same point in the soil are not possible.

The use of microlysimeters is also a gravimetric method (Boast & Robertson, 1982) that allows repeated measurements at the same time, for a direct estimate of soil evaporation rate in additions to soil water content. The procedure consists in inserting into the soil a small piece of aluminum or PVC pipe (10 to 20 cm in diameter and length). Then the pipe and the enclosed soil are removed by carefully excavating around the perimeter. The pipe is sealed on the bottom, weighed, then placed in a plastic bag and replaced in the same position in the soil, with the plastic bag pulled back to exposure the soil surface to the atmosphere. The soil surrounding the microlysimeter is repacked to resemble the original surface as closely as possible. At a later time the microlysimeter can be removed and reweighed to determine the water loss (soil surface evaporation) during the intervening time period. This may be done several times, after which the soil can be oven-dried and reweighed to back-calculate water content at each weighing. This is an inexpensive, direct and reasonably accurate measurement of soil evaporation (Lascano & van Bavel, 1986), but it is time-consuming and labor intensive. Since the soil in the core is not in hydraulic contact with the soil below, the evaporation rate form the core will eventually diverge from that of the surrounding soil, so a given core should not be used for more than a few days.

## 4. Neutron scattering

Neutron scattering is a time-tested indirect determination of soil water content. This method estimates the amount of water in a volume of soil by measuring the amount of hydrogen atoms present. A neutron probe consists of a source of fast or high energy neutrons and a detector, both housed in a unit which is lowered into an access tube installed in the soil. The probe is connected by a cable to a control unit located in the soil surface. Clips on the cable allow the cable to be set at pre-selected depths into the soil profile. Access tubes should be installed to the depth of the expected growth of the root crop. The control unit includes electronics for time control, a counter, memory and other electronics for processing readings (Figure 3).

This technique works based on the following principle. Fast neutrons emitted from the interaction of a radioactive alpha-emitter with Beryllium, pass through the access tube into the surrounding soil, where they gradually lose energy by collision with other atomic nuclei. Hydrogen atoms in the soil (mostly in water molecules) are effective in slowing the fast neutrons because they are of approximately the same mass. The result is a cloud of slow or thermalized neutrons; some of them diffuse back to the detector. The size and density of the cloud depends mainly on soil type and soil water content, and is spherical in shape (Figure 4) with a diameter of 15 to 40 cm. Thermalized neutrons that impact the detector create a small electrical impulse, which is amplified and counted. The number of slow neutrons counted in a specified interval of time is linearly related to the total volumetric soil water content. A higher count indicates higher soil water content.

Fig. 3. Neutron probe for measuring soil water content.

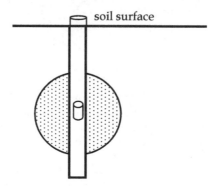

Fig. 4. Spatial sensitivity of neutron scattering in the soil.

Commercial neutron probes combine the source and detector in a single unit which fits in the access tube. They also include a standard material within the housing, so that a standard count may be taken prior to each measurement. This allows expressing the reading as count ratio (count in the soil/count in the standard), to account for changes in source strength associated with radioactive decay and for instrument drift.

Neutron probe must be calibrated for the soil type in which they will be used (Baker, 1990). Manufacturers provide a calibration curve with each neutron prove, but it is probably useful only for moisture measurements in homogeneous sands and gravels. Several studies have shown that factory-supplied curves give large errors when used in agricultural soils (Chanasky & McKenzie, 1986). Soil-specific calibration is necessary because detector readings are affected by the presence of non-water hydrogen (principally in organic matter), other elements in the soil with the ability to thermalized fast neutrons, and elements that absorb fast neutrons such as boron, cadmium and chlorine. The calibration procedure consists on compare neutron count ratios taken in a defined soil depth, against water content determined gravimetrically from samples taken nearby at the same soil depth.

The neutron probe allows relatively rapid and repeatable measurements of soil water content to be made at several depths and locations within a field. Repeatable measurements at the same location through the crop growing season, reduces the effect of soil variability on the measurements.

The main advantages of this method are: direct reading of soil water content, large volume of the soil is sampled, one unit can be used in several locations, and is accurate when properly calibrated. The main disadvantages are: individual calibration for each type of soil is required, difficult to use in automatic monitoring, its use near the surface requires spatial technique because of the escape of fast neutrons, and the high cost of the unit. There is also a radiation safety hazard, which requires special licensing, operation training, handling, shipping and storage procedures.

Example 1:

A homogeneous and deep soil has the following parameters: $\theta_{w\_FC}$ =0.285 g/g, $\theta_{w\_PWP}$ = 0.140 g/g, $\rho_b$ = 1.25 g cm$^{-3}$. The calibration equation of the neutron probe used to measure soil water content was: $\theta_w$ = -0.031 + 0.1496*C.R, where $\theta_w$ is the gravimetric water content (g/g) and C.R. is the counting ratio of the thermalized neutrons. If the neutron probe gave a reading of 1.452 in a soil depth of 40 cm, determine: depth of total available soil water (TAW), depth of depleted available soil water (DAW) and depth of residual available soil water (RAW). Assume that the density of water is 1 g/cm$^3$.

TAW is calculated using equation (6):

$$TAW = [\theta_{w\_FC} - \theta_{w\_PWP}]*(\rho b/\rho w)*Z$$

Substituting values in the above equation we get:

$$TAW = [0.285 - 0.140]*(1.25/1.0)*40 = 7.25 \text{ cm of water}$$

To calculate the depleted and residual available soil water, the soil water content at the time of measurement must be first calculated, using the calibrated equation of the neutron probe:

$$\theta_w = -0.031 + 0.1496*C.R$$

$$\theta_w = -0.031 + 0.1496*(1.452)$$

$$\theta_w = 0.186 \text{ g/g}$$

Similarly, DAW is calculated with equation (7):

$$DAW = [\theta_{w\_FC} - \theta_{w\_actual}]*(\rho b/\rho w)*Z$$

By substituting values we obtain:

$$DAW = [0.285 - 0.186]*(1.25/1.0)*40 = 4.95 \text{ cm of water}$$

RAW is calculated using equation (8)

$$RAW = [\theta_{w\_actual} - \theta_{w\_PWP}]*(\rho b/\rho w)*Z$$

Substituting values:

$$RAW = [0.186 - 0.140]*(1.25/1.0)*40 = 2.30 \text{ cm of water}$$

## 5. Tensiometers

Soil water tension, soil water suction or soil water potential are all terms describing the energy status of soil water. Soil water potential is a measure of the amount of energy with which water is held in the soil. A water release curve shows the relation between soil water content and soil water tension.

Tensiometers have been used for many years to measure soil water tension in the field. Tensiometers are water-filled tubes with a ceramic cup attached at one end and a vacuum gauge (or mercury manometer) airtight seal on the other end. The device is installed in the soil with the ceramic cup in good contact with the surrounding soil at the desired depth (Figure 5). The soil matric potential is measured by the vacuum gauge as water is pull out of the ceramic cup into the soil by matric forces. As the soil is rewetted, the tension gradient reduces and water flows into the ceramic cup. As the soil goes through wetting and drying cycles, tension readings can be taken.

Fig. 5. Use of tensiometers to determine soil matric potential at different soil depths.

Commercially available tensiometers use a vacuum gauge to read the tension in a scale from 0 to 100 kPa, although the practical operating range is from 0 to 70 kPa, because once air enters the tube, values are no longer accurate. If the water column is intact, a zero reading indicates saturated soil conditions. Readings of about 10 kPa correspond to field capacity for coarse-textured soils, while readings of around 30 kPa can approximate field capacity for fine-textured soils.

Tensiometer readings can be used as indicators of soil water content and the need for irrigation. When instruments installed at the active root zone of a given crop, reach a certain reading, they can be used to indicate when to start irrigation, based on soil texture and soil type. Similarly, instruments at deeper depths of the root zone may be used to indicate when adequate water has been applied. However, to determine the depth of water to applied, the curve that relates soil water content against soil water potential for the specific soil must be known.

Careful installation and maintenance of tensiometers is required for reliable results. The ceramic cup must be in intimate and complete contact with the soil. A few hours to a few days are required for the tensiometer to come to equilibrium with the surrounding soil. The tensiometer should be pumped with a hand vacuum pump to remove air bubbles. The length of the tensiometers is from 15 to 120 cm. It is recommended that the tensiometers be installed in pairs, one at 1/3 and the other at 2/3 of the crop rooting depth. They should be installed out of the way of traffic and cultivation. In freezing climates, insulate or remove tensiometers during winter months, because it takes only a small frost to knock the vacuum gauges out of calibration.

Tensiometers have been used to estimate water balance (Devitt *et al.*, 1983), follow capillarity rise above the water table (McIntyre, 1982) and characterize unsaturated soil hydraulic conductivity (Ward *et al.*, 1983). More recently, Zermeño-Gonzalez *et al.* (2007) used tensiometers to schedule irrigation in an orchard of lemon. They found that the highest fruit yield can be obtained when irrigation is applied at a reading of 30 kPa of tensiometers installed at a soil depth of 30 cm.

The main advantages of this method are: direct reading of soil water matric potential, inexpensive, automatic for continuous reading, relatively reliable. The main disadvantages are: requires the soil moisture characteristic curve to relate to soil water content, samples a small portion of soil near the cup may take a long time to reach equilibrium with the soil.

Example 2:

Zermeño-González *et al.* (2007) obtained a calibration equation to get soil moisture content as a function soil tension measured with a tensiometer installed at a soil depth of 30 cm. The equation was: L = 109.30 – 17.29*ln(Tens), where, L is the soil water content at a depth of 30 cm (mm/30 cm), Tens is the soil water tension (kPa). If the reading of the tensiometer was 40 kPa, determine the depth of water to be applied to take the soil water content to field capacity, assuming that for that soil and crop (an orchard of lemon) a soil water tension of 15 kPa corresponds to field capacity.

The depth of water to be applied to take the soil water content to field capacity can be calculated with the following relation:

$L_{\_to\_FC} = L_{15kPa} - L_{actual\_kPa}$ where: $L_{15kPa}$ is the soil water content at 15 Kpa (mm/30 cm) and $L_{actual\_kPa}$ is the soil water content that corresponds to the actual reading of the tensiometer (mm/30 cm). substituting the calibration equation in this relation we obtain:

$$L_{\_to\_FC} = [109.30 - 17.29*Ln(15)] - [109.30 - 17.229*Ln(40)]$$

$$L_{\_to\_FC} = [62.478 \text{ mm}] - [45.519 \text{ mm}]$$

$$L_{\_to\_FC} = 16.959 \text{ mm}/30 \text{ cm}$$

## 6. Porous blocks

Porous blocks are made of materials such as gypsum, ceramic, nylon and fiberglass. Similar to tensiometers, the blocks are buried in intimate contact with the soil at some desired depth and allowed to come to water tension equilibrium with the surrounding soil. Once equilibrium is reached, different properties of the block which are affected by water tension may be measured.

One of the more common types of porous blocks are electrical resistance blocks. Electrodes inside the block are used to measure the resistance to electrical current flow between them. In operation, measurements are made by connecting an ohmmeter to the electrodes of the resistance block. The resistance is proportional to the quantity of water in the block, which is a function of soil water tension. Higher resistance readings mean lower block water content and thus higher soil water tension. By contrast, lower resistance readings indicate higher block water content and lower soil water tension. A Useful technique is to calibrate blocks in soil on a pressure plate apparatus. In this way, resistance, water content and soil water potential can be determined simultaneously on each sample.

Resistance blocks work best in soils drier than -0.05 MPa, making the complementary in the range of operation to soil tensiometers. They are typically accurate to soil matric potentials as low as -2.0 to -3.0 MPa. Because response time of resistance blocks is slow, they are not useful for following rapid wetting events. Significant hysteresis effect may also be found between wetting and drying calibrations. Gypsum blocks require little maintenance and can be left in the field under frizzing conditions. Being made of gypsum, the block will slowly dissolve, requiring replacement. The rate of dissolution depends on soil pH and soil water conditions. Gypsum blocks are best suited for use in fine-textured soils. They are not sensitive to changes of soil water tension from 0 to 100 kPa. High soil salinity affects the electrical resistivity of the soil solution, although the gypsum buffers this effect to a certain degree.

Watermark blocks or granular matrix sensor, is a new style of electrical resistance block. The electrodes are embedded in a granular matrix material, similar to compressed fine sand. A gypsum wafer is embedded in the granular matrix near the electrodes. A synthetic porous membrane and a PVC casing with holes hold the block together. The granular matrix material enhances the movements of water to and from the surrounding soil, making the block more responsive to soil water tensions in the range from 0 to 100 kPa. These sensors have good sensitivity to soil water tension in a range of 0 to 200 kPa. This makes them more adaptable to a wide range of soil textures and irrigation regimes than gypsum blocks and tensiometers.

Readings are taken by attaching special electrical resistance meter to the wire leads and setting the estimated soil temperature. The readings of the Watermark meter are kPa of soil water tension, similar to the tensiometers. Watermark blocks require little maintenance and can be left in the soil under frizzing conditions. The blocks are much more stable and have a longer life than gypsum blocks. Soil salinity affects the electrical resistivity of the soil water solution and may cause erroneous readings. The gypsum wafer in the watermark blocks offers some buffering of this effect.

The main advantages of resistance blocks are: they are calibrated for soil water potential, are reliable, inexpensive, can be automated for monitoring. Disadvantages: requires the soil moisture characteristic curve to relate to water content, must be calibrated individually, and samples a small volume of soil.

Example 3:

At the agricultural experimental station of Universidad Autonoma Agraria Antono Narro, in Saltillo, Coahuila, Mexico, a Watermark block was calibrated against gravimetric measurements in a clay loam soil. The calibration was performed at a soil depth of 30 cm where the bulk density was 1.206 g cm³. Determine the depth of available soil water between 20 and 100 kPa, for a soil depth of 30 cm.

The calibration equation of the Watermark block was:

$$\theta w = 0.215 - 0.0005 * Tens$$

$$R^2 = 0.853$$

Where: $\theta w$ is the gravimetric water content $(g/g)$, Tens is the soil water tension (kPa).

The depth of available soil water (AW) between two gravimetric soil water contents can be calculated with the following equation:

$$AW = [\theta_{w1} - \theta_{w2}] * (\rho b / \rho w) * Z \tag{11}$$

where, $\theta_{w1}$ is the initial or higher gravimetric soil water content $(g/g)$, $\theta_{w2}$ is the final or lower gravimetric soil water content $(g/g)$ the other variables of equation (10) were previously defined. $\theta_{w1}$ and $\theta_{w2}$ are calculated by substituting 20 and 100 kPa respectively in the calibration equation of the Watermark block

$$\theta_{w1} = 0.215 - 0.0005 * Tens$$

$$\theta_{w1}1 = 0.215 - 0.0005 * (20)$$

$$\theta_{w1} = 0.205 \ g/g$$

$$\theta_{w2} = 0.215 - 0.0005 * Tens$$

$$\theta_{w2} = 0.215 - 0.0005 * (100)$$

$$\theta_{w2}2 = 0.165 \ g/g$$

Finally, substituting the value of : $\theta_{w1}$ and $\theta_{w2}$ in equation (10) the depth of available soil water is obtained:

$$AW = [\theta_{w1} - \theta_{w2}] * (\rho b / \rho w) * Z$$

$$AW = [0.205 - 0.165] * (1.206 / 1.00) * 30$$

$$AW = 1.447 \ cm; = 14.47 \ mm / 30 \ cm$$

## 7. Time domain reflectometry

Time-domain reflectometry (TDR) is a method for measuring soil water content, based in the determination of the dielectric permittivity of the porous media at microwave (MHz-

GHz) frequencies. The method uses equipment developed for testing coaxial cables in the telecommunications industry, which consists of a pulse generator, a sampler that produces a low frequency facsimile of high frequency signals, and an oscilloscope that displays the sampler output. Electromagnetic pulses of frequencies in the 1 MHz to 1 GHz region are sent down to a coaxial transmission line that ends in a parallel pair of stainless steel rods embedded in the soil. The unit samples and displays the reflected pulses, which exhibit perturbations at any point in the transmission line where impedance changes occur, as happens at the juncture of the cable with the steel waveguides. The termination of the transmission line at the end of the waveguides is also clearly visible on the oscilloscope since the remaining energy in the pulse is reflected at that point. The distance on the oscilloscope screen between these two points together with the known length of the waveguides allows calculation of the pulse propagation velocity (Vp), relative to the velocity of electromagnetic radiation in a vacuum ($c = 3*10^8$ m s$^{-1}$). From this relation the apparent dielectric permittivity (Ka) can be approximated by the equation:

$$Ka = \left(\frac{c}{vp}\right)^2 \tag{12}$$

The apparent dielectric permittivity of the soil depends on the volume fraction of the soil constituents and their respective dielectric permittivity. Ka of the dry minerals of the soils varies between 2 and 5, the air has a Ka of 1 while the Ka of water is approximately 80. This shows that Ka for the soil is strongly dependent on soil water content. Topp et al. (1980) found that a third order polynomial equation best fit the data between volumetric water content ($\theta v$) and the apparent dielectric permittivity of the soil (Ka), over the range of water content from air-dry to saturation.

$$\theta v = -5.3 * 10^{-2} + 2.92 * 10^{-2}Ka - 5.5 * 10^{-4}\,Ka^2 + 4.3 * 10^{-6}\,Ka^3 \tag{13}$$

Equations 12 and 13 show that the apparent dielectric permittivity of the soil is inversely related to the pulse propagation velocity, i.e., faster propagation velocity indicates a lower dielectric permittivity of the soil and thus lower soil water content. Or, as soil water content increases, propagation velocity decreases, and the dielectric permittivity of the soil increases.

Waveguides inserted into the soil consist of a pair of parallel stainless steel rods spaced between 3 and 5 cm apart. They can be installed in the soil horizontally, vertically at an angle of 45º etc. The TDR soil water measurement system measures the average volumetric soil water content along the length of the waveguide. The volume of soil sampled approximates a cylinder surrounding the waveguide with a diameter about 1.5 times the spacing of the parallel rods.

The waveguides may be permanently installed with wire leads brought to the surface, but this requires care to minimize soil disruption. Horizontal installation yields a depth-specific measurement, while insertion at a 45º angle integrates a larger volume of soil horizontally and vertically. Portable hand push waveguide probes can be used to measure at different locations in the upper soil profile which corresponds to the length of the waveguides. Waveguide must be carefully inserted into the soil with full soil contact along the entire length of the rods. Annular air gaps around the rods will affect readings of the low side. The waveguide rods must remain parallel when they are installed in the soil.

Once properly calibrated and installed, the TDR technique is highly accurate. Precise measurements may be made near the surface, which is an important advantage compare to other techniques such as the neutron probe. Research has shown (Evett *et al.*, 2001; Pedro-Vaz & Hopmans, 2001) that the dielectric permittivity of the soil is nearly independent of soil type and bulk density and relatively unaffected by soil salinity. Soil salinity or bulk electrical conductivity affects the degree of attenuation of electromagnetic pulse in the soil. Other studies (Jacobsen & Schjonning, 1993) found that inclusion of soil bulk density, clay and organic matter content in the calibration equation improves the correlation, suggesting that complex interactions between the soil components affect the electric properties of the soil.

The CS616 TDR probe (Campbell, Sci., Inc, USA) (Figure 6) consists of two stainless steel rods connected to a printed circuit board. A shielded four-conductor cable is connected to the circuit board to supply power, enable the probe, and monitor the pulse output. The circuit board is encapsulated in epoxy. High-speed electronic components on the circuit board are configured as a bistable multivibrator. The output of the multivibrator is connected to the probe rods which act as a waveguide.

The fundamental principle of CS616 operation is that an electromagnetic pulse will propagate along the probe rods at a velocity that is dependent on the dielectric permittivity of the material surrounding the rods. As water content increases, the propagation velocity decreases because polarization of water molecules takes time. The travel time of the applied signal along 2 times the rod length is essentially measured. The applied signal travels the length of the probe rods and is reflected from the rod ends traveling back to the probe head. A part of the circuit detects the reflection and triggers the next pulse. The Water Content Reflectometer output is essentially a square wave with an amplitude of +/- 0.7 volts and a period that fluctuates between 16 and 32 μs, which depends on the volumetric water content and is used for the calibration equation. For soil solution electrical conductivity values less than 2 dS m$^{-1}$ The calibration equation is: $\theta v = -0.0663-0.0063*t+0.0007*t^2$, where $\theta v$ is he volumetric soil water content ($m^3/m^3$) and t is the period of the square wave (μs).

Fig. 6. CS616 TDR probe for measurement of volumetric soil water content.

The main advantages of this method are: measures water content, samples large soil volume therefore decreases interference due to heterogeneity, can be automated for continuous readout, relatively stable over time. The main disadvantages are: Insertion of rods may be difficult, may sample excessively large soil volume, and requires the use of a datalogger.

Example 4:

A CS616 was used to measure the soil water content of the upper 30 cm of the soil profile in a soya bean crop. If the reading of the probe was 28 µs one day-after irrigation, and 25 µs seven days later, determine the crop evapotranspiration if no rain was observed during the TDR readings.

The volumetric water content one day after irrigation was:

$$\theta v\_1= -0.0663-0.0063*(28)+0.0007*(28)^2$$

$$\theta v\_1= 0.306 \ m^3/m^3$$

and 7 days later:

$$\theta v\_7= -0.0663-0.0063*(25)+0.0007*(25)^2$$

$$\theta v\_7= 0.214 \ m^3/m^3$$

The crop evapotranspiration (LamET) was the difference in volumetric water content during the seven days multiplied by the soil depth

$$LamET = (\theta v\_1 - \theta v\_7)*Soil\_depth$$

$$LamET = (0.306-0.214)*0.30$$

$$lamET = 0.0276 \ m$$

$$LamET = 27.6 \ mm$$

The average daily crop evapotranspiration (LamETprom) during the seven days was:

$$LamETprom = 27.6/7 = 3.943 \ mm$$

## 8. Impedance and capacitance methods

The Impedance and capacitance as well as the TDR techniques are electromagnetic (EM) sensors, which principle is based in the significant difference in the dielectric permittivity (Ka) between water, air and mineral particles of the soil. Therefore, is possible to establish a good relation between the soil water content ($m^3 \ m^{-3}$) and Ka, such as the Topp equation (Equation 12), (Topp et al., 1980).

EM sensors determine Ka of an unsaturated porous medium from different physical principles; transit time, impedance, capacitance, etc. For instance, the TDR (Time Domain Reflectometry) and TDT (Time Domain Transmission) techniques estimate Ka from the relationship between this and the transit time (ts) of an electromagnetic wave travelling along the rods of length L of a probe inserted into a porous medium, according to the following equation (Campbell, 1990):

$$Ka = \frac{(ts*c)^2}{(2*L)^2} \qquad (14)$$

where, c is the speed of light (m/s) in the vacuum.

Impedance sensors determine the amplitude difference in voltage due to changes in impedance, Z ($\Omega$), between the transmission line of the sensor and the rods that are inserted in the porous media, using the equation (Kelleners et al., 2005):

$$\sqrt{ka} = \frac{c*InvCotan(Z(\Omega))}{2*\pi*L} \qquad (15)$$

Capacitance methods, consider the composite media soil-probe as a capacitor whose capacitance, C (F), is proportional to Ka, according to the following equation:

$$C(F) = g(m)*Ka*Ko \qquad (16)$$

where, g(m) is a geometric factor and Ko =8.54 is the value of permittivity of the vacuum. The relation obtained between Ka or $\theta$ and the signal provided by a given EM sensor is known as the calibration equation. In general, the manufacturer of a specific EM sensor provides signal versus $\theta$ equations or signal versus Ka, valid for some conditions of media or soil type. However, because the soil is a heterogeneous porous medium of variable composition and since Ka depends on other variables such as the electrical conductivity of the medium or the frequency of the EM wave, It is recommended to perform a recalibration of the manufacturer equation of the sensor, especially when a more accurate determination of the soil water content is required.

Regalado et al. (2010) made a recalibration of the manufacturer equation of nine RM sensors. For the EC10 and EC20 capacitance probes of Decagan Devices, Inc, the manufacturer equations were:

$$\theta v = -0.376 + 9.36*10^{-4} *S \qquad (17)$$

and,

$$\theta v = -0.290 + 6.95*10^{-4}*S \qquad (18)$$

The ML2x impedance probe of Delata –T devices Ltd., the manufacturer equation was:

$$Ka^{0.5} = 1.07 + 6.40*10^{-3}*S - 6.40*10^{-6}*S^2 + 4.7*10^{-9}*S^3 \qquad (19)$$

where S is the reading signal of the sensor (mv).

After recalibration in a non saline solution of different values of dielectric permittivity (Ka), the new equations for the EC10 and EC20 capacitance probes were:

$$1/Ka = 0.0589/S^2 - 0.0455 \qquad (20)$$

$$1/Ka = -0.2581 + 0.0607*S + 0.2331/S \qquad (21)$$

And for the impedance probe was:

$$1/Ka = 0.134/S^{0.5} - 0.105 \tag{22}$$

They also concluded that after recalibration, all sensors behaved correctly under conditions equivalent to those of a non saline soil with sandy texture. Since the sensors studied performed acceptable for the entire range of water content, its suitability for a particular application should be decided according to other specific criteria such as volume of soil explored, robust probes, possibility of automation of the readings, cost, etc.

## 9. Conclusions

Understanding the soil water holding capacity and the factors affecting the plant available soil water are necessary for good Irrigation management. Adequate soil moisture is critical to plant growth. Too little water, or water applied at the wrong time, causes stress and reduces growth and too much may result in surface runoff, erosion and leaching of nutrients and pesticides.

Different techniques are currently available to directly measure or determine soil water content in a discrete or continuous manner. Some are very simple and others are more complex techniques. The cost of keeping track of soil water content is paid back through the benefits of effective water management, such as energy savings, water savings, water quality improvement, and improvement in quality and yield of harvest.

Successful implementation of any of the methods requires careful attention during the installation, operation, and maintenance of the equipment and sensors. Soil type, soil salinity and irrigation regime are important parameters that must be considered to choice a particular method or technique to get the best results. A routine sampling schedule should be implemented to obtain the most information from any of these methods. The difference in soil water content at a given location from one sampling time to the next often provides more information than random space and time measurements. Soil water should be measured or monitored in at least two depths in the active crop root zone at several locations in a field to obtain a field average.

There have been many advances in electromagnetic (EM) sensor technology (time domain reflectometry (TDR, impedance and capacitance-based approach) which have resulted in sensors that are more robust, less expensive, more suitable for different soil types that can be connected to advanced data loggers for a continuous monitoring of soil water content. Real-time, continuous measurement of soil moisture in the plant rooting zone is very important for determining crop evapotranspiration and the amount of water to apply.

## 10. References

Baker, J.M. (1990). Measurem Pearcy, R.W., Ehleringer, J., Mooney, H.A. & Rundel, P.W ent of soil water content. *Remote Sensing Reviews* 5(1):263-279.

Boast, C.W. & Robertson, T.M. (1982). A "micro-lysimeter" method for determining evaporation from bare soil: description and laboratory evaluation. *Soil Sci. Soc. Am. J.* 46:689-696.

Campbell, J.E. ( 1990). Dielectric properties and influence of conductivity in soils at one to fifty megahertz. *Soil Sci. Soc. Am. J.* 54: 332-341.

Chanasky, D.S. & McKenzie, R.H. (1986). Field calibration of a neutron probe. *Ca. J. Soil Sci.* 66:173-176.

Devitt, D., Jurry, W.A., Sternberg, P. & Stolzy, L.H. (1983). Comparison of methods used to estimate evapotranspiration for leaching control. *Irrig. Sci.* 4:59-69.

Evett, S. Laurent, J-P., Cepuder, P. & Hignett, C. (2001). Neutron scattering, capacitance, and water TDR soil water content measurements compared on four continents. *Proceedings of the 17th World Congress of soil Sci.*, 14-21 august, Thailand.

Gardner, W.H. (1986). Water Content. In: *Methods of Soil Analysis, Part 1. Amer. Soc. Agron.* Madison, WI. Pp 493-544.

Hanks, R.J. & Ashcroft, G.L. (1980). *Applied Soil Physics*, Springer-Verlag, Berlin, 159 p.

Hillel, D. (1980). Fundamentals of Soil Physics, Academic Press, new York, 413 p.

Jacobsen, O.H. & Schjonning. (1993). A laboratory calibration of time domain reflectometry for soil water measurement including effects of bulk density and texture. *Journal of Hydrology* 151(2-4):147-157.

Kelleners, T.J., Robinson, D.A., Shouse, P.J., Ayars, J.E. & Skaggs, T.H. (2005). Frequency dependence of the complex permittivity and its impact on dielectric sensor calibration in soils. *Soil Sci. Soc. Am. J.* 69: 67-76.

Klute, A. (1986). Water retention: laboratory methods . In: *Methods of Soil Analysis. Part 1. Physical and Mineralogical Methods*, 2nd edn, Agronomy Number 9 (Part 1), (ed. A. Klute), *American Society of Agronomy*, Madison,pp. 635-662, ISBN 0-89118-088-5

Lascano, R.J. & van Bavel, C.H.M. (1986). Simulation and measurement of evaporation from a bare soil. *Soil Sci. Soc. Am. J.* 50:1127-1133.

McIntyre, D.S. (1982). Capillary rise from saline groundwater in clay soil cores. *Aus. J. Soil. Res.*, 20: 305-313.

Regalado, C.M., Ritter, A. &Garcia, O. (2010). Dielectri response of commerial capacitance, impedance, and TDR electromagnetic sensors in standard liquid media, *proceedings of the third international symposium on soil water measurement using capacitance, impedance and TDT*, Murcia, Spain, april 2010.

Rundel, P.W. & Jarrel, W.M. (1991). *Water in the environment, In: Plant physiological ecology, field methods and instrumentation*, Pearcy, R.W., Ehleringer, J., Mooney, H.A. & Rundel, P.W, pp (29-56), Chapman and Hall, ISBN: 0-412-40730-2, London ,England.

Pedro-Vaz, C.M. & Hopmans, J.W.C. (2001). Simultaneous measurements of soil penetration resistance and water content with a combined penetrometer-TDR moisture probe, *Soil Sci. Soc. Am. J.* 65(1):4-12.

Topp, G.C., Davis, J.L. & Annan, A.P. (1980). Electromagnetic determination of soil water content: measurements in coaxial transmission lines. *Water Resour. Res.*16:574-582.

Ward, A., Wells, L.G. & Philips, R.E. (1983). Characterizing unsaturated hydraulic conductivity of Western Kentucky surface-mine spoils and soils. *Soil Sci. Soc. Am. J.* 47:847-854

Zermeño-González, A., García-Delgado, M.A., B.I. Castro-Meza, B.I. & Rodríguez-Rodríguez, H. (2007). Tensión de humedad del suelo y rendimiento de fruto en limón Italiano. *Revista Fitotecnia Mexicana*, 30(3): 295-302

# 12

# Comparison of Different Irrigation Methods Based on the Parametric Evaluation Approach in West North Ahwaz Plain

Mohammad Albaji[1], Saeed Boroomand Nasab[1] and Jabbar Hemadi [2]
*[1]Irrigation and Drainage Dept., Faculty of Water Sciences Eng.*
*Shahid Chamran University, Ahwaz*
*[2]Khuzestan Water & Power Authority, Ahwaz*
*Iran*

## 1. Introduction

Food security and stability in the world greatly depends on the management of natural resources. Due to the depletion of water resources and an increase in population, the extent of irrigated area per capita is declining and irrigated lands now produce 40% of the food supply (Hargreaves and Mekley.1998). Consequently, available water resources will not be able to meet various demands in the near future and this will inevitably result into the seeking of newer lands for irrigation in order to achieve sustainable global food security. Land suitability, by definition, is the natural capability of a given land to support a defined use. The process of land suitability classification is the appraisal and grouping of specific areas of land in terms of their suitability for a defined use.

According to FAO methodology (1976) land suitability is strongly related to "land qualities" including erosion resistance, water availability, and flood hazards which are in themselves immeasurable qualities. Since these qualities are derived from "land characteristics", such as slope angle and length, rainfall and soil texture which are measurable or estimable, it is advantageous to use the latter indicators in the land suitability studies, and then use the land parameters for determining the land suitability for irrigation purposes. Sys et al. (1991) suggested a parametric evaluation system for irrigation methods which was primarily based upon physical and chemical soil properties. In their proposed system, the factors affecting soil suitability for irrigation purposes can be subdivided into four groups:

- Physical properties determining the soil-water relationship in the soil such as permeability and available water content;
- Chemical properties interfering with the salinity/alkalinity status such as soluble salts and exchangeable Na;
- Drainage properties;
- Environmental factors such as slope.

Briza et al. (2001) applied a parametric system (Sys et al. 1991) to evaluate land suitability for both surface and drip irrigation in the Ben Slimane Province, Morocco, while no highly suitable areas were found in the studied area. The largest part of the agricultural areas was classified as marginally suitable, the most limiting factors being physical parameters such as slope, soil calcium carbonate, sandy soil texture and soil depth.

Bazzani and Incerti (2002) also provided a land suitability evaluation for surface and drip irrigation systems in the province of Larche, Morocco, by using parametric evaluation systems. The results showed a large difference between applying the two different evaluations. The area not suitable for surface irrigation was 29.22% of total surface and 9% with the drip irrigation while the suitable area was 19% versus 70%. Moreover, high suitability was extended on a surface of 3.29% in the former case and it became 38.96% in the latter. The main limiting factors were physical limitations such as the slope and sandy soil texture.

Bienvenue et al. (2003) evaluated the land suitability for surface (gravity) and drip (localized) irrigation in the Thies, Senegal, by using the parametric evaluation systems. Regarding surface irrigation, there was no area classified as highly suitable ($S_1$). Only 20.24% of the study area proved suitable ($S_2$, 7.73%) or slightly suitable ($S_3$, 12. 51%). Most of the study area (57.66%) was classified as unsuitable ($N_2$). The limiting factor to this kind of land use was mainly the soil drainage status and texture that was mostly sandy while surface irrigation generally requires heavier soils. For drip (localized) irrigation, a good portion (45.25%) of the area was suitable ($S_2$) while 25.03% was classified as highly suitable ($S_1$) and only a small portion was relatively suitable ($N_1$, 5 .83 %) or unsuitable ($N_2$, 5.83%). In the latter cases, the handicap was largely due to the shallow soil depth and incompatible texture as a result of a large amount of coarse gravel and/or poor drainage.

Mbodj et al. (2004) performed a land suitability evaluation for two types of irrigation i. e, surface irrigation and drip irrigation, in the Tunisian Oued Rmel Catchment using the suggested parametric evaluation. According to the results, the drip irrigation suitability gave more irrigable areas compared to the surface irrigation practice due to the topographic (slope), soil (depth and texture) and drainage limitations encountered with in the surface irrigation suitability evaluation.

Barberis and Minelli (2005) provided land suitability classification for both surface and drip irrigation methods in Shouyang county, Shanxi province, China where the study was carried out by a modified parametric system. The results indicated that due to the unusual morphology, the area suitability for the surface irrigation (34%) is smaller than the surface used for the drip irrigation (62%). The most limiting factors were physical parameters including slope and soil depth.

Dengize (2006) also compared different irrigation methods including surface and drip irrigation in the pilot fields of central research institute, Ikizce research farm located in southern Ankara. He concluded that the drip irrigation method increased the land suitability by 38% compared to the surface irrigation method. The most important limiting factors for surface irrigation in study area were soil salinity, drainage and soil texture, respectively whereas, the major limiting factors for drip or localized irrigation were soil salinity and drainage.

Liu et al. (2006) evaluated the land suitability for surface and drip irrigation in the Danling County, Sichuan province, China, using a Sys's parametric evaluation system. For surface irrigation the most suitable areas ($S_1$) represented about (24%) of Danling

County, (33%) was moderately suitable (S$_2$), (%9) was classified as marginally suitable (S$_3$), (7%) of the area was founded currently not suitable (N$_1$) and (25%) was very unsuitable for surface irrigation due to their high slope gradient. Drip irrigation was everywhere more suitable than surface irrigation due to the minor environmental impact that it caused. Areas highly suitable for this practice covered 38% of Danling County; about 10% was marginally suitable (the steep dip slope and the structural rolling rises of the Jurassic period). The steeper zones of the study area (23%) were either approximately or totally unsuitable for such a practice.

Albaji et al. (2007) carried out a land suitability evaluation for surface and drip Irrigation in the Shavoor Plain, in Iran. The results showed that 41% of the area was suitable for surface irrigation ;50% of the area was highly recommend for drip irrigation and the rest of the area was not considered suitable for either irrigation method due to soil salinity and drainage problem.

Albaji et al. (2010a) compared the suitability of land for surface and drip irrigation methods according to a parametric evaluation system in the plains west of the city of Shush, in the southwest Iran. The results indicated that a larger amount of the land (30,100 ha − 71.8%) can be classified as more suitable for drip irrigation than surface irrigation.

Albaji et al. (2010b) investigated different irrigation methods based upon a parametric evaluation system in an area of 29,300 ha in the Abbas plain located in the Elam province, in the West of Iran. The results demonstrated that by applying sprinkler irrigation instead of surface and drip irrigation methods, the arability of 21,250 ha (72.53%) in the Abbas plain will improve.

Albaji et al. (2010c) also provided a land suitability evaluation for surface, sprinkle and drip irrigation systems in Dosalegh plain: Iran. The comparison of the different types of irrigation techniques revealed that the drip and sprinkler irrigations methods were more effective and efficient than that of surface irrigation for improved land productivity. However, the main limiting factor in using either surface or/and sprinkler irrigation methods in this area were soil texture, salinity, and slope, and the main limiting factor in using drip irrigation methods were the calcium carbonate content, soil texture and salinity.

Albaji and Hemadi (2011) evaluated the land suitability for different irrigation systems based on the parametric evaluation approach on the Dasht Bozorg Plain:Iran. The results showed that by applying sprinkle irrigation instead of drip and surface irrigation, the arability of 1611.6 ha (52.5%) on the Dasht Bozorg Plain will improve. In addition, by applying drip irrigation instead of sprinkle or surface irrigation, the land suitability of 802.4 ha (26.2%) on this plain will improve. Comparisons of the different types of irrigation systems revealed that sprinkle and drip irrigation were more effective and efficient than surface irrigation for improving land productivity. It is noteworthy, however, that the main limiting factor in using sprinkle and/or drip irrigation in this area is the soil calcium carbonate content and the main limiting factors in using surface irrigation are soil calcium carbonate content together with drainage.

The main objective of this research is to evaluate and compare land suitability for surface, sprinkle and drip irrigation methods based on the parametric evaluation systems for the West North Ahwaz Plain, in the Khuzestan Province, Iran.

## 2. Materials and methods

The present study was conducted in an area about 37324.91 hectares in the West north ahwaz Plain, in the Khuzestan Province, located in the West of Iran during 2009-2011. The study area is located 5 km West north of the city of Ahwaz, $31^{\circ} 20'$ to $31^{\circ} 40'$ N and $48^{\circ} 36'$ to $48^{\circ} 47'$ E. The Average annual temperature and precipitation for the period of 1965-2004 were 24.5 $C^{\circ}$ and 210 mm, respectively. Also, the annual evaporation of the area is 2,550 mm (Khuzestan Water & Power Authority [KWPA], 2005). The Karun River supplies the bulk of the water demands of the region. The application of irrigated agriculture has been common in the study area. Currently, the irrigation systems used by farmlands in the region are furrow irrigation, basin irrigation and border irrigation schemes.

The area is composed of two distinct physiographic features i.e. River Alluvial Plains and Plateaux, of which the River Alluvial Plains physiographic unit is the dominating features. Also, twenty two different soil series were found in the area (Table.1).

The semi-detailed soil survey report of the West north ahwaz plain (KWPA. 2009) was used in order to determine the soil characteristics. Table.2 has shown some of physico – chemical characteristics for reference profiles of different soil series in the plain. The land evaluation was determined based upon topography and soil characteristics of the region. The topographic characteristics included slope and soil properties such as soil texture, depth, salinity, drainage and calcium carbonate content were taken into account. Soil properties such as cation exchange capacity (CEC), percentage of basic saturation (PBC), organic mater (OM) and pH were considered in terms of soil fertility. Sys et al. (1991) suggested that soil characteristics such as OM and PBS do not require any evaluation in arid regions whereas clay CEC rate usually exceeds the plant requirement without further limitation, thus, fertility properties can be excluded from land evaluation if it is done for the purpose of irrigation.

Based upon the profile description and laboratory analysis, the groups of soils that had similar properties and were located in a same physiographic unit, were categorized as soil series and were taxonomied to form a soil family as per the Keys to Soil Taxonomy (2008). Ultimately, twenty two soil series were selected for the surface, sprinkle and drip irrigation land suitability.

In order to obtain the average soil texture, salinity and $CaCo_3$ for the upper 150cm of soil surface, the profile was subdivided into 6 equal sections and weighting factors of 2, 1.5, 1, 0.75, 0.50 and 0.25 were used for each section, respectively (Sys et al.1991).

For the evaluation of land suitability for surface, sprinkle and drip irrigation, the parametric evaluation system was used (Sys et al. 1991). This method is based on morphology, physical and chemical properties of soil.

Six parameters including slope, drainage properties, electrical conductivity of soil solution, calcium carbonates status, soil texture and soil depth were also considered and rates were assigned to each as per the related tables, thus, the capability index for irrigation (Ci) was developed as shown in the equation (1):

$$Ci = A \times \frac{B}{100} \times \frac{C}{100} \times \frac{D}{100} \times \frac{E}{100} \times \frac{F}{100} \tag{1}$$

where A, B, C, D, E, and F are soil texture rating, soil depth rating, calcium carbonate content rating, electrical conductivity rating, drainage rating and slope rating, respectively.

| Series No | Characteristics description |
|---|---|
| 1 | Soil texture " Heavy : *CL", without salinity and alkalinity limitation, Depth 150 cm, level to very gently sloping : 0 to 2%, imperfectly drained. |
| 2 | Soil texture " Heavy : CL", very severe salinity and alkalinity limitation, Depth 100 cm, level to very gently sloping : 0 to 2%, poorly drained. |
| 3 | Soil texture " Medium : SL", without salinity and alkalinity limitation, Depth 150 cm, level to very gently sloping : 0 to 2%, moderately drained. |
| 4 | Soil texture " Heavy : SIC", without salinity and alkalinity limitation, Depth 120 cm, level to very gently sloping : 0 to 2%, imperfectly drained. |
| 5 | Soil texture " Medium : SL", without salinity and alkalinity limitation, Depth 150 cm, level to very gently sloping : 0 to 2%, moderately drained. |
| 6 | Soil texture " Very Heavy: C", slight salinity and alkalinity limitation, Depth 125 cm, level to very gently sloping : 0 to 2%, poorly drained. |
| 7 | Soil texture " Very Heavy : SIC", very severe salinity and alkalinity limitation , Depth 140cm, level to very gently sloping : 0 to 2%, very poorly drained. |
| 8 | Soil texture" Very Heavy: C", severe salinity and alkalinity limitation, Depth 150cm, level to very gently sloping : 0 to 2%, very poorly drained. |
| 9 | Soil texture" Heavy: SICL", without salinity and alkalinity limitation, Depth 110 cm, level to very gently sloping : 0 to 2%, poorly drained. |
| 10 | Soil texture" Very Heavy: C", severe salinity and alkalinity limitation, Depth 150 cm, level to very gently sloping : 0 to 2%, very poorly drained. |
| 11 | Soil texture " Very Heavy : C", very severe salinity and alkalinity limitation, Depth 110 cm, level to very gently sloping : 0 to 2%, very poorly drained. |
| 12 | Soil texture " Medium : L", without salinity and alkalinity limitation, Depth 170 cm, level to very gently sloping : 0 to 2%, moderately drained. |
| 13 | Soil texture " Heavy : SICL", without salinity and alkalinity limitation, Depth 150 cm, level to very gently sloping : 0 to 2%, well drained. |
| 14 | Soil texture " Very Heavy : SIC", moderate salinity and alkalinity limitation, Depth 150 cm, level to very gently sloping : 0 to 2%, imperfectly drained. |
| 15 | Soil texture " Heavy: SICL", without salinity and alkalinity limitation, Depth 135 cm, level to very gently sloping : 0 to 2%, well drained. |
| 16 | Soil texture" Heavy: SICL",very without salinity and alkalinity limitation, Depth 150cm, level to very gently sloping : 0 to 2%, well drained. |
| 17 | Soil texture" Heavy: SCL", without salinity and alkalinity limitation, Depth 150 cm, level to very gently sloping : 0 to 2%, well drained. |
| 18 | Soil texture " Medium: SIL", slight salinity and alkalinity limitation, Depth 135 cm, level to very gently sloping : 0 to 2%, well drained. |
| 19 | Soil texture " Heavy : SICL", without salinity and alkalinity limitation , Depth 140cm, level to very gently sloping : 0 to 2%, well drained. |
| 20 | Soil texture" Heavy: SICL", without salinity and alkalinity limitation, Depth 150cm, level to very gently sloping : 0 to 2%, well drained. |
| 21 | Soil texture " Medium : SIL", slight salinity and alkalinity limitation, Depth 140 cm, level to very gently sloping : 0 to 2%, well drained. |
| 22 | Soil texture " Heavy : SICL", slight salinity and alkalinity limitation, Depth 130 cm, level to very gently sloping : 0 to 2%, well drained. |

* Texture symbols: LS: Loamy Sand, SL: Sandy Loam, L: Loam , SIL: Silty Loam , CL: Clay Loam , SICL: Silty Clay Loam , SCL: Sandy Clay Loam , SC: Sandy Clay , SIC: Silty Clay , C: Clay.

Table 1. Soil series of the study area.

| Soil seris.No | Soil seris.name | Depth (Cm) | Soil texture | ECe (ds.m-1) | pH | OM (%) | CEC (meq/100g) | CaCo3 (%) |
|---|---|---|---|---|---|---|---|---|
| 1 | Veyss | 150 | CL | 1.50 | 7.90 | 0.24 | 8.54 | 48.00 |
| 2 | Omel Gharib | 100 | CL | 48.00 | 7.70 | 0.46 | 5.61 | 49.00 |
| 3 | Ramin | 150 | SL | 1.10 | 7.80 | 0.39 | 8.19 | 41.00 |
| 4 | Amerabad | 120 | SIC | 3.50 | 8.50 | 0.23 | 10.31 | 48.00 |
| 5 | Solieh | 150 | SL | 3.40 | 7.90 | 0.29 | 5.57 | 34.00 |
| 6 | Band Ghir | 125 | C | 4.10 | 8.00 | 0.52 | 15.24 | 35.00 |
| 7 | Abu Baghal | 140 | SIC | 52.00 | 8.10 | 0.37 | 11.43 | 45.00 |
| 8 | Sheykh Mussa | 150 | C | 17.50 | 8.40 | 0.56 | 13.26 | 46.00 |
| 9 | Safak | 110 | SICL | 3.90 | 8.10 | 0.47 | 13.53 | 40.00 |
| 10 | Molla Sani | 150 | C | 21.50 | 7.90 | 0.36 | 12.91 | 39.00 |
| 11 | Teal Bomeh | 110 | C | 55.00 | 7.90 | 0.68 | 9.85 | 49.00 |
| 12 | Karkheh | 170 | L | 2.70 | 7.70 | 0.29 | 6.49 | 46.00 |
| 13 | Karun 1 | 150 | SICL | 2.20 | 7.70 | 0.25 | 9.21 | 47.00 |
| 14 | Shoteyt | 150 | SIC | 9.50 | 7.90 | 0.60 | 8.66 | 47.00 |
| 15 | Abbasieh 1 | 140 | SICL | 1.10 | 7.60 | 0.39 | 8.63 | 51.00 |
| 16 | Deylam 1 | 150 | SICL | 2.90 | 7.50 | 0.28 | 10.48 | 50.00 |
| 17 | Qalimeh | 150 | SCL | 1.20 | 7.90 | 0.26 | 12.05 | 49.00 |
| 18 | Abbasieh 2 | 135 | SIL | 5.90 | 7.60 | 0.39 | 12.73 | 44.00 |
| 19 | Karun 2 | 140 | SICL | 1.00 | 7.60 | 0.41 | 10.22 | 51.00 |
| 20 | Deylam 2 | 150 | SICL | 3.40 | 7.50 | 0.32 | 10.81 | 49.00 |
| 21 | Ghaleh Nasir | 140 | SIL | 4.20 | 7.60 | 0.38 | 11.56 | 51.00 |
| 22 | Abdul Amir | 130 | SICL | 7.50 | 7.80 | 0.57 | 10.38 | 46.00 |

Table 2. Some of physico – chemical characteristics for reference profiles of different soil series.

In Table 3 the ranges of capability index and the corresponding suitability classes are shown.

| Capability Index | Definition | Symbol |
|---|---|---|
| > 80 | Highly Suitable | S1 |
| 60-80 | Moderately Suitable | S2 |
| 45-59 | Marginally Suitable | S3 |
| 30-44 | Currently Not Suitable | N1 |
| < 29 | Permanently Not Suitable | N2 |

Table 3. Suitability Classes for the Irrigation Capability Indices (Ci) Classes.

In order to develop land suitability maps for different irrigation methods (Figs.2-5), a semi-detailed soil map (Fig.1) prepared by Albaji was used, and all the data for soil characteristics were analyzed and incorporated in the map using ArcGIS 9.2 software.

The digital soil map base preparation was the first step towards the presentation of a GIS module for land suitability maps for different irrigation systems. The Soil map was then digitized and a database prepared. A total of twenty two different polygons or land mapping units (LMU) were determined in the base map. Soil characteristics were also given for each LMU. These values were used to generate the land suitability maps for surface, sprinkle and drip irrigation systems using Geographic Information Systems.

## 3. Results and discussion

Over much of the West north ahwaz Plain, the use of surface irrigation systems has been applied specifically for field crops to meet the water demand of both summer and winter crops .The major irrigated broad-acre crops grown in this area are wheat, barley, and maize, in addition to fruits , melons, watermelons and vegetables such as tomatoes and cucumbers. There are very few instances of sprinkle and drip irrigation on large area farms in the West north ahwaz Plain.

Twenty two soil series and eighty six series phases or land units were derived from the semi-detailed soil study of the area(Table.1).  The land units are shown in Fig.1 as the basis for further land evaluation practice. The soils of the area are of Aridisols and Entisols orders. Also, the soil moisture regime is Aridic and Aquic while the soil temperature regime is Hyperthermic (KWPA.2003).

**Legend**

**Series**

| | | | |
|---|---|---|---|
| River | | 11 | |
| 1 | | 12 | |
| 2 | | 13 | |
| 3 | | 14 | |
| 4 | | 15 | |
| 5 | | 16 | |
| 6 | | 17 | |
| 7 | | 18 | |
| 8 | | 19 | |
| 9 | | 20 | |
| 10 | | 21 | |
| | | 22 | |

0 1.5 3      6      9      12
Kilometers

Fig. 1. Soil Map of the Study Area.

As shown in Tables 4 and 5 for surface irrigation, the soil series coded 13, 16, 17, 18 and 20 (4233.46 ha - 11.36%) were highly suitable ($S_1$); soil series coded 1, 12, 15, 19, 21 and 22 (14041.96 ha - 37.62 %) were classified as moderately suitable ($S_2$), soil series coded 3, 4, 5 and 9 (8835.99 ha - 23.66%) were found to be marginally suitable ($S_3$). soil series coded 6 and 14 (1033.86 ha - 2.77%) were classified as currently not-suitable ($N_1$) and soil series coded 2, 7, 8, 10 and 11 (8714.66 ha - 23.34 %) were classified as permanently not-suitable ($N_2$)  for any surface irrigation practices.

The analysis of the suitability irrigation maps for surface irrigation (Fig. 2) indicate that some portion of the cultivated area in this plain (located in the east) is deemed as being highly suitable land due to deep soil, good drainage, texture, salinity and proper slope of the area. The moderately suitable area is mainly located to the center, and east of this area due to soil texture and drainage limitations. Other factors such as depth and slope have no influence on the suitability of the area whatsoever. The map also indicates that some part of the cultivated area in this plain was evaluated as marginally suitable because of the

| Codes of Land Units | Surface Irrigation | | Sprinkle Irrigation | | Drip Irrigation | |
|---|---|---|---|---|---|---|
| | Ci | suitability classes | Ci | suitability classes | Ci | suitability classes |
| 1 | 70.2 | S2 sw [a] | 76.5 | S2 sw [b] | 72 | S2 sw [c] |
| 2 | 11.40 | N2 snw | 12.6 | N2 snw | 12.8 | N2 snw |
| 3 | 59.23 | S3 sw | 76.95 | S2 s | 76 | S2 s |
| 4 | 52.21 | S3 sw | 57.37 | S3 sw | 54.4 | S3 sw |
| 5 | 59.23 | S3 sw | 76.95 | S2 s | 76 | S2 s |
| 6 | 40.27 | N1snw | 47.23 | S3 sw | 45.22 | S3 sw |
| 7 | 17.90 | N2 snw | 22.37 | N2 snw | 22.1 | N2 snw |
| 8 | 20.88 | N2 snw | 25.81 | N2 snw | 25.5 | N2 snw |
| 9 | 52.65 | S3 sw | 58.5 | S3 sw | 56 | S3 sw |
| 10 | 20.88 | N2 snw | 25.81 | N2 snw | 25.5 | N2 snw |
| 11 | 17.90 | N2 snw | 22.37 | N2 snw | 22.1 | N2 snw |
| 12 | 71.07 | S2 sw | 76.95 | S2 s | 72 | S2 S |
| 13 | 87.75 | S1 | 90 | S1 | 80 | S1 |
| 14 | 41.76 | N1snw | 48.76 | S3 snw | 46.24 | S3 snw |
| 15 | 78 | S2 s | 80 | S1 | 70 | S2 S |
| 16 | 87.75 | S1 | 90 | S1 | 80 | S1 |
| 17 | 83.36 | S1 | 85.5 | S1 | 76 | S2 S |
| 18 | 83.36 | S1 | 85.5 | S1 | 76 | S2 S |
| 19 | 78 | S2 S | 80 | S1 | 70 | S2 S |
| 20 | 87.75 | S1 | 90 | S1 | 80 | S1 |
| 21 | 74.1 | S2 s | 76 | S2 S | 66.5 | S2 S |
| 22 | 78.97 | S2 sn | 85.5 | S1 | 76 | S2 S |

a & b . Limiting Factors for Surface and Sprinkle Irrigations: n: (Salinity & Alkalinity), w: (Drainage) and s: (Soil Texture).
c. Limiting Factors for Drip Irrigation: s: (Calcium Carbonate & Soil Texture), w: (Drainage)  and n: (Salinity & Alkalinity).

Table 4. Ci Values and Suitability Classes of Surface ,Sprinkle and Drip irrigation for Each Land Units.

drainage and soil texture limitations. The current non-suitable land and permanently non-suitable land can be observed only in the west and center of the plain because of very severe limitation of salinity & alkalinity, drainage and soil texture. For almost the total study area elements such as soil depth, slope and $CaCO_3$ were not considered as limiting factors.

In order to verify the possible effects of different management practices, the land suitability for sprinkle and drip irrigation was evaluated (Tables 4 and 5).

For sprinkle irrigation, soil series coded 13, 15, 16, 17, 18, 19, 20 and 22 (9329.14 ha – 25.01%) were highly suitable ($S_1$) while soil series coded 1, 3, 5, 12 and 21 (14938.7 ha- 40.02%) were classified as moderately suitable ($S_2$). Further, soil series coded 4, 6, 9 and 14 (3877.43 ha – 10.38%) were found to be marginally suitable ($S_3$) and soil series coded 2, 7, 8, 10 and 11 (8714.66 ha – 23.34 %) were classified as permanently not-suitable ($N_2$) for sprinkle irrigation.

| Suitability | Surface Irrigation | | | Sprinkle Irrigation | | | Drip Irrigation | | |
|---|---|---|---|---|---|---|---|---|---|
| | Land unit | Area (ha) | Ratio (%) | Land unit | Area (ha) | Ratio (%) | Land unit | Area (ha) | Ratio (%) |
| S1 | 13 , 16 , 17 , 18 , 20 | 4233.46 | 11.36 | 13 , 15 , 16 , 17 , 18 , 19 20 , 22 | 9329.14 | 25.01 | 13 , 16 , 20 | 1724.88 | 4.64 |
| S2 | 1 , 12 , 15 , 19 , 21 , 22 | 14041.96 | 37.62 | 1 , 3 , 5 , 12 , 21 | 14938.7 | 40.02 | 1 , 3 , 5 , 12 , 15 , 17 , 18 , 19 , 21 , 22 | 22542.96 | 60.39 |
| S3 | 3 , 4 , 5 , 9 | 8835.99 | 23.66 | 4 , 6 , 9 , 14 | 3877.43 | 10.38 | 4 , 6 , 9 , 14 | 3877.43 | 10.38 |
| N1 | 6 , 14 | 1033.86 | 2.77 | - | - | - | - | - | - |
| N2 | 2 , 7 , 8 , 10 , 11 | 8714.66 | 23.34 | 2 , 7 , 8 , 10, 11 | 8714.66 | 23.34 | 2 , 7 , 8 , 10, 11 | 8714.66 | 23.34 |
| [a]Mis Land | | 464.99 | 1.25 | | 464.99 | 1.25 | | 464.99 | 1.25 |
| Total | | 37324.91 | 100 | | 37324.91 | 100 | | 37324.91 | 100 |

a. Miscellaneous Land: (Hill, Sand Dune and River Bed)

Table 5. Distribution of Surface, Sprinkle and Drip Irrigation Suitability.

Regarding sprinkler irrigation, (Fig. 3) the highly suitable area can be observed in the some part of the cultivated zone in this plain (located in the east) due to deep soil, good drainage, texture, salinity and proper slope of the area. As seen from the map, the largest part of the cultivated area in this plain was evaluated as moderately suitable for sprinkle irrigation because of the moderate limitations of drainage and soil texture. Other factors such as depth, salinity and slope never influence the suitability of the area. The marginally suitable lands are located only in the North and south of the plain. The permanently non-suitable land can be observed in the west and center of the plain and their non-suitability of the land are due to the severe limitations of salinity & alkalinity, drainage and soil texture. The current non-suitable lands did not exist in this plain. For almost the entire study area slope, soil depth and $CaCO_3$ were never taken as limiting factors.

Fig. 2. Land Suitability Map for Surface Irrigation.

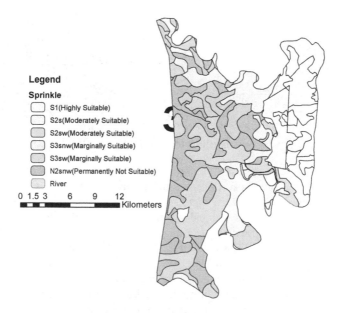

Fig. 3. Land Suitability Map for Sprinkle Irrigation.

For drip irrigation, soil series coded 13, 16 and 20 (1724.88 ha-4.64%) were highly suitable (S$_1$) while soil series coded 1, 3, 5, 12, 15, 17, 18, 19, 21 and 22 (22542.96 ha- 60.39%) were classified as moderately suitable (S$_2$). Further, soil series coded 4, 6, 9 and 14 (3877.43 ha, 10.38%) were found to be slightly suitable (S$_3$) and soil series coded 2, 7, 8, 10 and 11 (8714.66 ha – 23.34 %) were classified as permanently not-suitable (N$_2$) for drip irrigation.

Regarding drip irrigation, (Fig. 4) the highly suitable lands covered the smallest part of the plain. The slope, soil texture, soil depth, calcium carbonate, salinity and drainage were in good conditions .The moderately suitable lands could be observed over a large portion of the plain (east, north and south parts) due to the medium content of calcium carbonate. The marginally suitable lands were found only in the Northwest and southeast of the area .The limiting factors for this land unit were drainage and the medium content of calcium carbonate. The permanently non-suitable land can be observed in the west and center of the plain and their non-suitability of the land are due to the severe limitations of calcium carbonate, salinity & alkalinity, drainage and soil texture. The current non-suitable lands did not exist in this plain. For almost the entire study area slope, soil depths were never taken as limiting factors.

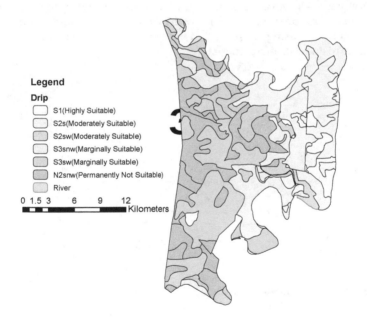

**Legend**

**Drip**

☐ S1(Highly Suitable)

☐ S2s(Moderately Suitable)

☐ S2sw(Moderately Suitable)

☐ S3snw(Marginally Suitable)

☐ S3sw(Marginally Suitable)

☐ N2snw(Permanently Not Suitable)

☐ River

0 1.5 3    6    9    12 Kilometers

Fig. 4. Land Suitability Map for Drip Irrigation.

The mean capability index (Ci) for surface irrigation was 55.90 (Marginally suitable) while for sprinkle irrigation it was 62.33 (Moderately suitable). Moreover, for drip irrigation it was 58.31 (Marginally suitable). For the comparison of the capability indices for surface, sprinkle and drip irrigation. Tables 6 indicated that in soil series coded 2 applying drip irrigation systems was the most suitable option as compared to surface and sprinkle irrigation systems. In soil series coded 1,3,4,5,6,7,8,9,10,11,12,13,14,15,16,17,18,19,20,21 and 22 applying sprinkle irrigation systems was more suitable then surface and drip irrigation systems. Fig.5 shows the most suitable map for surface, sprinkle and drip irrigation systems in the West north ahwaz plain as per the capability index (Ci) for different irrigation systems. As seen from this map, the largest part of this plain was suitable for sprinkle irrigation systems and some parts of this area was suitable for drip irrigation systems.

The results of Tables 4, 5 and 6indicated that by applying sprinkle irrigation instead of surface and drip irrigation methods, the land suitability of 35038,81 ha (93.87%) of the west north ahwaz Plain's land could be improved substantially. However by applying drip Irrigation instead of surface and sprinkle irrigation methods, the suitability of 1821,12 ha (4.88%) of this Plain's land could be improved. The comparison of the different types of irrigation revealed that sprinkle irrigation was more effective and efficient then the drip and surface irrigation methods and improved land suitability for irrigation purposes. The second best option was the application of drip irrigation which was considered as being more practical than the surface irrigation method. To sum up the most suitable irrigation systems for the west north ahwaz Plain' were sprinkle irrigation, drip irrigation and surface irrigation respectively. Moreover, the main limiting factor in using surface and sprinkle irrigation methods in this area were salinity & alkalinity, drainage and soil texture and the main limiting factors in using drip irrigation methods were the salinity & alkalinity, drainage, soil texture and calcium carbonate.

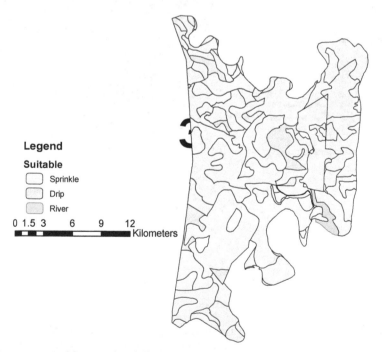

Fig. 5. The most suitable map for different irrigation systems.

| Codes of Land Units | The Maximum Capability Index for Irrigation(Ci) | Suitability Classes | The Most Suitable Irrigation Systems | Limiting Factors |
|---|---|---|---|---|
| 1 | 76.5 | S2 sw | Sprinkle | Soil Texture and Drainage |
| 2 | 12.8 | N2 snw | Drip | CaCo3& Soil Texture, Salinity & Alkalinity and Drainage |
| 3 | 76.95 | S2 s | Sprinkle | Soil Texture |
| 4 | 57.37 | S3 sw | Sprinkle | Soil Texture and Drainage |
| 5 | 76.95 | S2 s | Sprinkle | Soil Texture |
| 6 | 47.23 | S3 sw | Sprinkle | Soil Texture and Drainage |
| 7 | 22.37 | N2 snw | Sprinkle | Soil Texture , Salinity & Alkalinity and Drainage |
| 8 | 25.81 | N2 snw | Sprinkle | Soil Texture , Salinity & Alkalinity and Drainage |
| 9 | 58.5 | S3 sw | Sprinkle | Soil Texture and Drainage |
| 10 | 25.81 | N2 snw | Sprinkle | Soil Texture , Salinity & Alkalinity and Drainage |
| 11 | 22.37 | N2 snw | Sprinkle | Soil Texture , Salinity & Alkalinity and Drainage |
| 12 | 76.95 | S2 s | Sprinkle | Soil Texture |
| 13 | 90 | S1 | Sprinkle | No Exist |
| 14 | 48.76 | S3 snw | Sprinkle | Soil Texture , Salinity & Alkalinity and Drainage |
| 15 | 80 | S1 | Sprinkle | No Exist |
| 16 | 90 | S1 | Sprinkle | No Exist |
| 17 | 85.5 | S1 | Sprinkle | No Exist |
| 18 | 85.5 | S1 | Sprinkle | No Exist |
| 19 | 80 | S1 | Sprinkle | No Exist |
| 20 | 90 | S1 | Sprinkle | No Exist |
| 21 | 76 | S2s | Sprinkle | Soil Texture |
| 22 | 85.5 | S1 | Sprinkle | No Exist |

Table 6. The Most Suitable Land Units for Surface, Sprinkle and Drip Irrigation Systems by Notation to Capability Index (Ci) for Different   Irrigation Systems.

## 4. Conclusions

Several parameters were used for the analysis of the field data in order to compare the suitability of different irrigation systems. The analyzed parameters included soil and land characteristics. The results obtained showed that sprinkle and drip irrigation systems are more suitable than surface irrigation method for most of the study area. The major limiting factor for both sprinkle and surface irrigation methods were salinity & alkalinity, drainage and soil texture. However for drip irrigation method, salinity & alkalinity, drainage, soil texture and calcium carbonate were restricting factors. The results of the comparison between the maps indicated that the introduction of a different irrigation management policy would provide an optimal solution in as such that the application of sprinkle and drip irrigation techniques could provide beneficial and advantageous. This is the current strategy adopted by large companies cultivating in the area and it will provide to be economically viable for Farmers in the long run.Such a change in irrigation management practices would imply the availability of larger initial capitals to farmers (different credit conditions, for example) as well as a different storage and market organization. On the other hand, because of the insufficiency of water in arid and semi arid climate, the optimization of water use efficiency is necessary to produce more crops per drop and to help resolve water shortage problems in the local agricultural sector. The shift from surface irrigation to high-tech irrigation technologies, e.g. sprinkle and drip irrigation systems, therefore, offers significant water-saving potentials. On the other hand, since sprinkle and drip irrigation systems typically apply lesser amounts of water (as compared with surface irrigations methods) on a frequent basis to maintain soil water near field capacity, it would be more beneficial to use sprinkle and drip irrigations methods in this plain.

In this study, an attempt has been made to analyze and compare three irrigation systems by taking into account various soil and land characteristics. The results obtained showed that sprinkle and drip irrigation methods are more suitable than surface or gravity irrigation method for most of the soils tested. Moreover, because of the insufficiency of surface and ground water resources, and the aridity and semi-aridity of the climate in this area, sprinkle and drip irrigation methods are highly recommended for a sustainable use of this natural resource; hence, the changing of current irrigation methods from gravity (surface) to pressurized (sprinkle and drip) in the study area are proposed.

## 5. Acknowledgements

The writers gratefully acknowledge the Research and Standards Office for Irrigation and Drainage Networks of Khuzestan Water and Power Authority (KWPA) and Shahid Chamran University for their financial support and assistance during the study and field visits.

## 6. References

Albaji, M. Landi, A. Boroomand Nasab, S. & Moravej, K.(2007). Land Suitability Evaluation for Surface and Drip Irrigation in Shavoor Plain Iran *Journal of Applied Sciences,* 8(4):654-659.

Albaji, M. Boroomand Nasab, S. Kashkoli, H.A. & Naseri, A.A. (2010a). Comparison of different irrigation methods based on the parametric evaluation approach in the plain west of Shush, Iran. *Irrig. & Drain.* ICID.

Albaji, M. Boroomand Nasab, S. Naseri, A.A. & Jafari, S. (2010b). Comparison of different irrigation methods based on the parametric evaluation approach in Abbas plain— Iran. *J. Irrig. and Drain Eng. (ASCE)* 136 (2), 131–136.

Albaji, M. Shahnazari,A. Behzad,M. Naseri, A.A. Boroomand Nasab, S. & Golabi,M.(2010c). Comparison of different irrigation methods based on the parametric evaluation approach in Dosalegh plain: Iran. *Agricultural Water Management* 97 (2010) 1093– 1098.

Albaji, M. & Hemadi, J. (2011). Investigation of different irrigation systems based on the parametric evaluation approach on the Dasht Bozorg Plain. *Transactions of the Royal Society of South Africa.* Vol. 66(3), October 2011.DOI: 10.1080/0035919X.2011.622809. http://www.informaworld.com.

Barberis, A. & Minelli, S. (2005). *Land Evaluation in the Shouyang County, Shanxi Province, China.* 25th Course Professional Master. 8th Nov 2004-23 Jun 2005.IAO,Florence,Italy.
< http://www.iao.florence.it/training/geomatics/Shouyang/China_25hq.pdf>.

Bazzani, F. & Incerti, F. (2002). *Land Evaluation in the Province of Larache,Morocco.* 22nd Course Professional Master. Geometric and Natural Resources Evaluation. 12 Nov 2001-21 June 2002. IAO, Florence, Italy.    < http:// www.iao.florence.it/training/geomatics/Larache/Morocco_22.pdf>.

Bienvenue, J .S. Ngardeta, M. & Mamadou, K. (2003). *Land Evaluation in the Province of Thies, Senegal* . 23rd Course Professional Master. Geometric and Natural Resources Evaluation.8th Nov 2002-20 June 2003. IAO, Florence,Italy.
<http://www.iao.florence.it/training/geomatics/Thies/senegal23.pdf>.

Briza, Y. Dileonardo, F. & Spisni, A. (2001). *Land Evaluation in the Province of Ben Slimane, Morocco.* 21 st Course Professional Master.Remote Sensing and Natural Resource Evaluation .10 Nov 2000 - 22 June 2001. IAO, Florence, Italy.
<http://www.iao.florence.it/training/geomatics/BenSlimane/Marocco21.pdf>.

Dengize, O. (2006). A Comparison of Different Irrigation Methods Based on the Parametric Evaluation Approach. *Turk. J. Agric For,* 30: 21 – 29.
http://journals.tubitak.gov.tr/agriculture/issues/tar-06-30-1/tar-30-1-3-0505- 15.pdf.

Food and Agriculture Organization of the United Nations (FAO). (1976). A Framework for Land Evaluation. *Soil Bulletin No.32.* FAO,Rome,Italy.72 pp.
< http:// www.fao.org/docrep/x5310e/x5310e00.htm>.

Hargreaves, H.G. & Mekley, G.P.   (1998). *Irrigation fundamentals.* Water Resource Publication, LLC. 200 Pp.

Keys to soil taxonomy. (2008). U.S, Department of Agriculture. By Soil Survey Staff. Washington, DC, USA.< http://www.statlab.iastate.edu/soils/keytax>.

Khuzestan Water and Power Authority (KWPA). (2005*). Meteorology Report of West north ahwaz Plain, Iran* (in Persian)."< http://www.kwpa.com>.

Khuzestan Water and Power Authority (KWPA). (2009). *Semi-detailed Soil Study Report of West north ahwaz Plain, Iran* (in Persian). <http://www.kwpa.com>.

Liu, W., Qin, Y. & Vital, L. (2006). *Land Evaluation in Danling county ,Sichuan province, China.* 26th Course Professional Master. Geometric and Natural Resources Evaluation. 7th Nov 2005–23rd Jun 2006. IAO, Florence, Italy.
< http://www.iao.florence.it/training/geomatics/Danling/China_26.pdf>.

Mbodj ,C. Mahjoub, I. & Sghaiev, N.( 2004). *Land Evaluation in the Oud Rmel Catchment, Tunisia.* 24th Course Professional Master. Geometric and Natural Resources Evaluation. 10th Nov 2003–23 rd Jun 2004. IAO, Florence, Italy.
<http:// www.iao.florence.it/training/geomatics/Zaghouan/Tunisia_24lq.pdf>.

Sys, C. Van Ranst, E . & Debaveye, J. (1991). *Land Evaluation, Part l, Principles in Land Evaluation and Crop Production Calculations.* International Training Centre for Postgraduate Soil Scientists, University Ghent.
< http://www.plr.ugent.be/publicatie.html>.

# Permissions

The contributors of this book come from diverse backgrounds, making this book a truly international effort. This book will bring forth new frontiers with its revolutionizing research information and detailed analysis of the nascent developments around the world.

We would like to thank Dr. Manish Kumar, for lending his expertise to make the book truly unique. He has played a crucial role in the development of this book. Without his invaluable contribution this book wouldn't have been possible. He has made vital efforts to compile up to date information on the varied aspects of this subject to make this book a valuable addition to the collection of many professionals and students.

This book was conceptualized with the vision of imparting up-to-date information and advanced data in this field. To ensure the same, a matchless editorial board was set up. Every individual on the board went through rigorous rounds of assessment to prove their worth. After which they invested a large part of their time researching and compiling the most relevant data for our readers. Conferences and sessions were held from time to time between the editorial board and the contributing authors to present the data in the most comprehensible form. The editorial team has worked tirelessly to provide valuable and valid information to help people across the globe.

Every chapter published in this book has been scrutinized by our experts. Their significance has been extensively debated. The topics covered herein carry significant findings which will fuel the growth of the discipline. They may even be implemented as practical applications or may be referred to as a beginning point for another development. Chapters in this book were first published by InTech; hereby published with permission under the Creative Commons Attribution License or equivalent.

The editorial board has been involved in producing this book since its inception. They have spent rigorous hours researching and exploring the diverse topics which have resulted in the successful publishing of this book. They have passed on their knowledge of decades through this book. To expedite this challenging task, the publisher supported the team at every step. A small team of assistant editors was also appointed to further simplify the editing procedure and attain best results for the readers.

Our editorial team has been hand-picked from every corner of the world. Their multi-ethnicity adds dynamic inputs to the discussions which result in innovative outcomes. These outcomes are then further discussed with the researchers and contributors who give their valuable feedback and opinion regarding the same. The feedback is then collaborated with the researches and they are edited in a comprehensive manner to aid the understanding of the subject.

Apart from the editorial board, the designing team has also invested a significant amount of their time in understanding the subject and creating the most relevant covers. They scrutinized every image to scout for the most suitable representation of the subject and create an appropriate cover for the book.

The publishing team has been involved in this book since its early stages. They were actively engaged in every process, be it collecting the data, connecting with the contributors or procuring relevant information. The team has been an ardent support to the editorial, designing and production team. Their endless efforts to recruit the best for this project, has resulted in the accomplishment of this book. They are a veteran in the field of academics and their pool of knowledge is as vast as their experience in printing. Their expertise and guidance has proved useful at every step. Their uncompromising quality standards have made this book an exceptional effort. Their encouragement from time to time has been an inspiration for everyone.

The publisher and the editorial board hope that this book will prove to be a valuable piece of knowledge for researchers, students, practitioners and scholars across the globe.

# List of Contributors

Gabriele Dono and Luca Giraldo
University of Tuscia, Viterbo, Italy

Kenneth Nhundu and Abbyssinia Mushunje
University of Fort Hare, South Africa

Anindita Sarkar
Department of Geography, Miranda House, Delhi University, City, Delhi, India

M. A. Fernández-Zamudio
Valencian Institute of Agricultural Researchs, Spain

F. Alcon and M. D. De-Miguel
Technical University of Cartagena, Spain

Hang Zheng and Zhongjing Wang
State Key Laboratory of Hydro-Science and Engineering, Department of Hydraulic and Hydropower Engineering Tsinghua University, Beijing, China

Roger Calow
Overseas Development Institute, London, UK

Yongping Wei
Australia-China Water Resource Research Center, Department of Civil and Environmental Engineering, University of Melbourne, Melbourne, Victoria, Australia

Nasima Tanveer Chowdhury
Environmental Economics Unit (EEU), Department of Economics, Gothenburg University, Gothenburg, Sweden

Tim Aus der Beek
Department of Geography, Heidelberg University, Germany

Ellen Kynast and Martina Flörke
Center for Environmental Systems Research, University of Kassel, Germany

Kazunori Takahashi, Jan Igel and Holger Preetz
Leibniz Institute for Applied Geophysics, Germany

Seiichiro Kuroda
National Institute for Rural Engineering, Japan

**John D. Lea-Cox**
Professor, Department of Plant Science and Landscape Architecture, University of Maryland, College Park, USA

**Kristoph-Dietrich Kinzli**
Florida Gulf Coast University, USA

**Alejandro Zermeño-González , Martín Cadena-Zapata, Santos Gabriel Campos-Magaña and Raúl Rodríguez-García**
Antonio Narro Autonomous Agrarian University, México

**Juan Munguia-López and Luis Ibarra- Jiménez**
Department of Agricultural Plastics, Research Center for Applied Chemistry, Saltillo, Coahuila, México

**Mohammad Albaji and Saeed Boroomand Nasab**
Irrigation and Drainage Dept., Faculty of Water Sciences Eng. Shahid Chamran University, Ahwaz, Iran

**Jabbar Hemadi**
Khuzestan Water & Power Authority, Ahwaz, Iran